Genetic and Evolutionary Computation

Series Editors

The Genetic and Evolutionary Computation Book Series publishes research monographs, edited collections, and graduate-level texts in one of the most exciting areas of Computer Science. As researchers and practitioners alike turn increasingly to search, optimization, and machine-learning methods based on mimicking natural evolution to solve problems across the spectrum of the human endeavor, this growing field will continue to surprise with novel applications and results. Recent award-winning PhD theses, special topics books, workshops and conference proceedings in the areas of EC and Artificial Life Studies are of interest.

Areas of coverage include applications, theoretical foundations, technique extensions and implementation issues of all areas of genetic and evolutionary computation. Topics may include, but are not limited to: Optimization (multi-objective, multi-level), Design, control, classification, and system identification, Data mining and data analytics, Pattern recognition and deep learning, Evolution in machine learning, Evolvable systems of all types, Automatic programming and genetic improvement.

The GEVO book series (https://link.springer.com/series/7373) is the result of a merger the two former book series: Genetic Algorithms and Evolutionary Computation https://www.springer.com/series/6008 and Genetic Programming https://www.springer.com/series/6016.

Somnath Mukhopadhyay · Sunita Sarkar ·
Kalyanmoy Deb · Paramartha Dutta
Editors

Contemporary Advancement in Evolutionary Multi-objective Optimization

Springer

Editors
Somnath Mukhopadhyay
Department of Computer Science
and Engineering
Assam University
Silchar, Assam, India

Sunita Sarkar
Department of Computer Science
and Engineering
Assam University
Silchar, Assam, India

Kalyanmoy Deb
School of Engineering and Computer
Science
Michigan State University
East Lansing, MI, USA

Paramartha Dutta
Department of Computer and System
Science
Visva Bharati University
Santiniketan, West Bengal, India

ISSN 1932-0167 ISSN 1932-0175 (electronic)
Genetic and Evolutionary Computation
ISBN 978-981-95-7907-5 ISBN 978-981-95-7908-2 (eBook)
https://doi.org/10.1007/978-981-95-7908-2

This Springer imprint is published by the registered company Springer Nature Singapore Pte Ltd.
The registered company address is: 152 Beach Road, #21-01/04 Gateway East, Singapore 189721, Singapore

Contents

Balancing Exploration-Exploitation Using Switching Mechanism Integrated with Adaptive Heterogeneous Comprehensive Learning PSO and L-SHADE

Somnath Mukhopadhyay and Sunita Sarkar

Abstract This chapter proposes a method for improving the balance between exploration and exploitation in optimization algorithms. This method is particularly useful for combining algorithms to achieve a seamless exploration-exploitation equilibrium. The proposed hybrid algorithm combines two well-known optimization methodologies: Linear Population Size Reduction (L-Shade), which is known for its excellent convergence properties, and Adaptive Heterogeneous Comprehensive Particle Swarm Optimization (AHPSO), which is capable of both exploration and exploitation. The hybrid algorithm capitalizes on the strengths of each constituent algorithm via a meticulously designed workflow, ensuring robust performance across a wide range of optimization tasks. The algorithm adapts flexibly to changing optimization landscapes by combining adaptive mechanisms and Dynamic Phase Switching (DPS), demonstrating versatility and efficiency. The performance of the proposed algorithm is verified by ten benchmark functions consisting of 2 unimodal functions, 3 basic functions, 2 hybrid functions and 3 composition functions in the CEC2020 Competition on Single Objective Bound Constrained Optimization. The proposed algorithm is compared with multi-population L-Shade (mlL-SHADE) and L-Shade. The results show the competitiveness of the proposed algorithm.

Keywords AHPSO · Exploration · Exploitation · Dynamic phase switching · L-Shade

1 Introduction

For quite some time, the problem of complex optimization has been a subject of intense research. Researchers seek algorithms that are simple to implement, reliable, user-friendly, and effective in solving the given problem. The Metaheuristic opti-

S. Mukhopadhyay (✉) · S. Sarkar
Department of Computer Science and Engineering,, Assam University, Silchcar, India
e-mail: som.cse@live.com

S. Mukhopadhyay et al. (eds.), *Contemporary Advancement in Evolutionary Multi-objective Optimization*, Genetic and Evolutionary Computation,
https://doi.org/10.1007/978-981-95-7908-2_1

mization techniques have experienced significant progress and application in several fields over the past decade, including engineering [19, 21], healthcare industry [6, 17], etc. These methods are highly regarded for their ability to enhance the speed at which convergence occurs. Among these are those inspired by biological principles, especially natural selection, which have received a huge following and success. These fall under the category of evolutionary algorithms, which involve a wide range of approaches in which a set of potential solutions evolves iteratively through generations to reach the best solution possible such as Particle Swarm Optimization (PSO) [13], Heterogeneous Comprehensive Learning PSO (HCLPSO) [10], etc. The latter includes methods such as Differential Evolution (DE) [11], SHADE [14] among others, with L-SHADE [15] being the most recent advancement within this class.

Two key concepts, exploration and exploitation, form the basis of population-based algorithms. Exploration refers to the ability to investigate various areas within the global search space that contain the most favorable answers. Exploitation refers to the capacity to concentrate on certain local areas that have been found through exploration in order to achieve the most up-to-date ideal solutions [20].

Metaheuristics excel at efficiently finding and attaining globally optimum solutions by effectively managing the simultaneous processes of exploration and exploitation. Metaheuristic optimization has demonstrated its efficacy in several sectors like finance, logistics and supply chain management, telecommunications, manufacturing, energy, transportation, bioinformatics, environmental management, robotics and automation, and economics. These businesses leverage the capacity of metaheuristics to offer optimal solutions to complex and practical problems.

The primary issue lies in optimizing these algorithms to achieve efficiency without premature convergence. It happens when an optimization algorithm quickly settles on a suboptimal solution, missing better ones. Of course, it is important to mention notable approaches such as Heterogeneous Comprehensive Learning Particle Swarm Optimization (HCLPSO) [10] plus its adaptive versions and Heterogeneous Particle Swarm Optimization (HPSO) that are trying to reconcile these two aspects, although they still have some drawbacks. These algorithms either fail to cope with high-dimensional multi-modal problems or require further improvement in order to take advantage of their exploration and exploitation abilities [3, 4].

The selection of suitable metaheuristics for hybridization is a critical step that necessitates careful consideration, as it helps to maintain a harmonious equilibrium between exploration and exploitation, as well as amplifies the vulnerabilities of each. The hybridization of the Adaptive Heterogeneous Comprehensive Learning Particle Swarm Optimization (AHPSO) algorithm and L-Shade in this study aims to obtain a global optimal solution that does not veer off towards local optima in both unimodal and multimodal test functions.

The objective of this chapter is to come up with a new hybrid optimization method that combines the benefits of AHPSO, which allows for more exploration, and L-SHADE, which focuses on exploitation. The goal of this strategy is to achieve a state of dynamic equilibrium between these two opposing forces in order to improve algorithms' efficiency in intricate optimization scenarios.

The contribution of the chapter are:

- Introduced an innovative method to maintain an equilibrium between exploration and exploitation.
- Developed a dynamic performance-based phase-switching mechanism to transition between algorithms when they become ensnared in a local minima.
- Integrated the dynamic phase switching mechanism to develop the modified Adaptive Heterogeneous Comprehensive Particle Swarm Optimization (mAHPSO) algorithm to address the issue of becoming trapped in local minima for hybridization.
- Integrated the dynamic phase switching mechanism to developed the modified L-SHADE (mLSHADE) algorithm to address the issue of becoming trapped in local minima for hybridization.
- We developed a hybrid algorithm by integrating modified AHPSO (mAHPSO) and L-SHADE (mLSHADE) into hL-AHPSO-SM. This algorithm combines mAHPSO and mLSHADE with a switching mechanism.
- Examined the advantages and disadvantages of the proposed method and its potential for application in real-world optimization challenges.

We structured the subsequent sections of this chapter as follows: Sect. 2 presents a thorough examination of the existing literature, emphasizing the shortcomings of the current study and the rationale for the research's objectives. Section 3 outlines the techniques utilized in our hybrid algorithm. Section 4 showcases the results, along with the necessary examinations and comparisons. Section 5 presents a conclusion of the chapter and future suggestions.

2 Literature Review

Metaheuristics are a new group of complex algorithms that are being created to solve hard optimization problems that are continuous, discrete, unbounded, and have more than one goal [1]. These are more comprehensive heuristics that improve the ability to explore and exploit [20]. Metaheuristic techniques are classified into a variety of categories, such as physics-based, human-based, evolutionary-based, and swarm-based [5]. Swarm-based metaheuristics are derived from the behaviors of animals and vegetation that possess the capacity to interact with one another and their environment.

The focus of ongoing research has been on improving optimization algorithms using a variety of techniques. Many Metaheuristic Algorithms, such as PSO and DE, find widespread application in continuous, discrete, constrained, and unconstrained optimization problems. Presently, the main objective of the study is to enhance optimization algorithms by employing a diverse array of techniques. Metaheuristic methods, such as Particle Swarm Optimization (PSO) [9] and Differential Evolution (DE) [12], are commonly used to solve optimization issues. We can categorize these problems as continuous, discrete, constrained, or unconstrained. In 2021, an advanced hybrid meta-heuristic algorithm called haDEPSO [18], which integrates advanced Differential Evolution (aDE) and advanced Particle Swarm Optimization (aPSO) algorithms, was proposed. The work aims to address the limitations of conventional optimization methods and achieve better solutions for complex optimization problems by leveraging the strengths of both DE and PSO, while also making advancements in each algorithm to enhance its individual strengths. This algorithm was tested on 23 basic functions, 30 CEC 2014, and 30 CEC 2017 benchmark test suites, which demonstrated superior accuracy and convergence speed compared to other state-of-the-art algorithms. In addition, a hybrid Particle Swarm Optimization (PSO) method combines extensive learning and dynamic multi-swarm tactics. It divides the population into subpopulations for exploration and exploitation purposes. The approach also employs a self-regulating nonlinear inertia weight with Cauchy mutation to avoid converging to local optima. Nevertheless, MLPSO's heightened computational intricacy and challenges in execution are possible disadvantages.In addition, a hybrid Particle Swarm Optimization (PSO) method incorporates extensive learning and dynamic multi-swarm tactics. It divides the population into subpopulations for exploration and exploitation purposes. Furthermore, it utilizes a self-regulating nonlinear inertia weight with Cauchy mutation to avoid convergence to local optima. Nevertheless, MLPSO's heightened computational intricacy and challenges in execution are possible disadvantages [22].

The Top-Level Dual Exploitation PSO (TLDEPSO) algorithm utilizes gene editing technology and local neighborhood search to improve the optimization process of particle swarm optimization (PSO). The TLDEPSO algorithm exhibits limits in its ability to handle issues with a high number of dimensions and does not surpass the performance of leading algorithms from the CEC competition across all benchmarks [7]. As a result of the problems with older PSO algorithms, a new type of optimization called Adaptive Heterogeneous Comprehensive Learning Particle Swarm Optimization (AHPSO) [23] has been created to fix those problems. Similarly, a new variant of DE, called L-SHADE [15], has been developed to enhance the performance of DE algorithms.

The AHPSO algorithm categorizes the population into two distinct subpopulations, emphasizing the processes of exploitation and exploration. The utilization of an Adaptive Inertia Weight (AIW) approach by the exploitation subpopulation enhances the precision of local searches by leveraging historical data. The exploratory subpopulation improves the ability to search globally and maintains diversity through the use of an adaptive dimension mutation (ADM) technique and Dynamic Opposite Learning (DOL). To avoid the algorithm being stuck in local minima and to achieve a faster convergence rate, this dynamic phase-switching technique allows it to swap phases as needed.

The results of AHPSO demonstrate outstanding potential. It achieves ideal results on a considerable number of test functions in both 30-D and 50-D conditions, indicating its remarkable performance on multimodal functions. AHPSO specializes in solving high-dimensional multimodal issues and composition functions with exceptional accuracy and performance. The algorithm's convergence properties suggest that it can significantly speed up the convergence process and avoid becoming stuck in local optima.

Nevertheless, AHPSO does have limitations. It is not always reliable in finding the best possible solution globally, and it shows uneven performance on certain complex and multimodal functions. Real-time applications are not well-suited for AHPSO because of their low time complexity. Furthermore, theoretical alignment of exploration and exploitation can enhance the algorithm's robustness and stability.

L-SHADE is a more advanced version of the Differential Evolution technique that uses Linear Population Size Reduction (LPSR) to improve the optimization process's performance. L-SHADE extends the SHADE [14] methodology by introducing LPSR, a technique that optimizes while linearly decreasing the population size. This approach simplifies the population reduction process by requiring only the initial population size as input.

The L-SHADE algorithm continues to be highly relevant in global optimization and metaheuristic search algorithms. Its popularity stems from its ability to efficiently solve complex optimization problems, including those in constrained real-world engineering contexts. The L-SHADE algorithm has seen various advancements and adaptations to enhance its performance for specific optimization problems. Some notable variants of L-SHADE include: iL-SHADE [2], mL-SHADE [8], mL-SHADE [24], [16] etc.

The application of L-SHADE produces notable results. It outperforms SHADE 1.1 and other Differential Evolution (DE) algorithms in a variety of dimensions (10, 30, 50, and 100). L-SHADE excels at optimizing complex configurations, particularly composite functions in which subsets of variables exhibit different features. The system's simplicity, defined by a small number of control parameters, makes it easier to use and more effective at reducing population size while maintaining excellent performance.

Nevertheless, L-SHADE has some drawbacks. The complexities of the situation determine the relative effectiveness of LPSR and DPSR. L-SHADE's stability worsened as the dimensions exceeded 100, as opposed to when the dimensions were less. Furthermore, while reducing the number of components may be necessary, changing these configurations can impact the LPSR's overall effectiveness. This emphasizes the importance of correct calibration for a variety of issues.

The AHPSO and L-SHADE algorithms have made significant contributions to metaheuristics research by providing robust and efficient solutions to complex optimization problems. Nonetheless, all algorithms face the same challenge: finding an optimal balance between exploration and exploitation. Despite using approaches to improve both local and global search capabilities, AHPSO occasionally has difficulty determining the global optimum. Similarly, L-SHADE requires careful calibration to ensure balance, despite its effectiveness in reducing population size. To maximize the exploration of new options while also implementing proven solutions, future research should prioritize improving and refining these algorithms. These adjustments will result in advantages in adaptability, durability, and efficacy in a broad range of optimization conditions.

Despite appreciable developments in the current optimization algorithms, remarkable gaps still exist. No algorithm can find the optimal balance between exploration and exploitation, which is important for effective performance, especially on complex and high-dimensional problems. AHPSO also tends to show very inconsistent performance over some of the benchmark functions, while L-SHADE efficiency decreases as dimensionality increases, thus necessitating the development of more robust and reliable algorithms. Although the time complexity makes AHPSO less suitable for real-time applications, the calibration of the parameters in both algorithms constitutes a delicate issue and acts as a barrier to their applicability on a larger scale. To achieve a better balance between local and global search, we can combine AHPSO's adaptive phase-switching mechanisms with the population size reduction strategies of L-SHADE. Such a hybridization approach will underline the strengths of each algorithm, hence making the resultant one more robust, efficient, and adaptable in obtaining effective solutions for a much larger range of complex optimization problems. We compared the proposed model with state-of-the-art algorithms based on the results of the CEC 2020 benchmark test [8], which shows the algorithm's superiority over the other conventional algorithms.

The motivation of this research is simplified below:

- To mitigate the conventional algorithms drawback successfully by hybridization.
- Implement an adaptive feedback mechanism to change learning methods dynamically.
- To develop a method with balanced exploration and exploitation capabilities to solve both unimodal and multimodal problems efficiently
- To ensure a faster convergence rate by adopting the L-SHADE algorithm technique in hybridization.
- To ensure the developed models are validated and evaluated with the comparisons of existing hybrid algorithms to have a solid justification for the performance of the proposed method.

3 Methodology

Since hybridization consists of two conventional algorithms merging to form a technique, this section will briefly describe the modification, flow, structure, characteristics, and mathematical modeling of both the traditional AHPSO and LSHADE algorithms. Additionally, this section will provide a brief workflow of the proposed hybrid algorithm (hL-AHPSO-SM).

Hybrid metaheuristics have emerged as a crucial method for tackling optimization issues by achieving a harmonious equilibrium between exploration and exploitation. Furthermore, a hybrid metaheuristic also mitigates the constraints of traditional algorithms, as both algorithms aim to nullify each other's shortcomings

This section is organized as follows:

- **Dynamic Phase Switching**
- **Modified AHPSO (mAHPSO)**
- **Modified LSHADE (mLSHADE)**
- **Proposed Hybrid L-SHADE and AHPSO using Adaptive Switching Mechanism (hL-AHPSO-SM)**

3.1 Dynamic Phase Switching

The Dynamic Phase Switching technique is implemented with the Adaptive Heterogeneous Comprehensive Particle Swarm Optimization (AHPSO) algorithm and L-SHADE algorithm. It prevent the algorithm from getting stuck in local minima. It achieves this when the flag variable reaches a predefined threshold (referred as T_max), it indicates that the algorithm is likely stuck in a local minimum. The *Flag* variable is used to update the count when the proposed model fails to show any significant differences in the results.

The progression of the *Flag* variable is depicted in Eq. (1):

$$\text{Flag} = \begin{cases} \text{Flag} + 1 & \text{if gbest\_prev} = \text{gbest} \\ 0 & \text{otherwise} \end{cases} \tag{1}$$

where

- gbest: represents the current global best solution.
- gbest_prev: represents the previous global best solution.

The update of gbest_prev is shown in Eq. (2):

$$\text{gbest\_prev} = \begin{cases} \text{gbest\_prev} & \text{if gbest\_prev} = \text{gbest} \\ \text{gbest} & \text{otherwise} \end{cases} \tag{2}$$

The *Flag* variable is checked against T_max, and if they are equal, the algorithm returns, indicating a phase switch as shown in Eq. (3):

$$\text{Algorithm's state} = \begin{cases} \text{Switch algorithm} & \text{if Flag} == \text{T\_max} \\ \text{continue} & \text{otherwise} \end{cases} \tag{3}$$

By incorporating this dynamic phase switching mechanism, the algorithm enhances its robustness and ability to navigate complex, multimodal optimization landscapes.

3.2 Modified AHPSO (mAHPSO)

The modified AHPSO incorporates the use of Dynamic Phase Switching, further enhancing its optimization capabilities. Dynamic Phase Switching allows the algorithm to transition between different search strategies based on the current optimization state. This mechanism is crucial for maintaining a balance between exploration and exploitation, as it helps the algorithm adapt to different stages of the search process. By dynamically adjusting the search strategy, AHPSO can prevent premature convergence, adapt to varying optimization challenges, and enhance its efficiency by allocating computational resources more effectively. This adaptability makes the modified AHPSO robust and versatile, capable of achieving superior performance in solving complex and diverse optimization problems.

AHPSO also enhances PSO by introducing adaptive strategies and population diversity mechanisms [23]. The population is divided into two subpopulations: exploitation and exploration. Each subpopulation is managed with different strategies to maintain diversity and enhance performance.

- Exploitation Subpopulation: Utilizes an Adaptive Inertia Weight (AIW) method based on historical information of particles to refine local search capabilities. The inertia weight adapts according to each particle's fitness, improving convergence towards optimal solutions.
- Exploration Subpopulation: Implements an Adaptive Dimension Mutation (ADM) strategy to enhance the algorithm's ability to explore the search space and solve multimodal problems. Dynamic-Opposite Learning (DOL) is also applied to maintain diversity and avoid premature convergence. The exploration subpopulation does not rely on information from the exploitation subpopulation, ensuring independent and diverse search paths.

Algorithm (1) represents the detailed mechanism of the mAHPSO algorithm outlining the initialization steps, population division, and dynamic adaptation strategies employed in each subpopulation. The algorithm initializes with the setup of parameters, including population sizes, acceleration coefficients, inertia weights, and maximum iterations. The highlighted part of the algorithm represents the incorporation of Dynamic Phase Switching, which helps the algorithm avoid getting stuck at local minima and reduces the chance of premature convergence

3.3 Modified LSHADE (mLSHADE)

The modified LSHADE (mLSHADE) algorithm also incorporates the use of dynamic phase switching to further enhance its optimization capabilities. Equations (1), (2) and (3) represents the use of Dynamic Phase Switching. Incorporating dynamic phase switching into mLSHADE adds a layer of adaptability and robustness to the optimization process. This enhancement ensures that the algorithm can dynamically

Algorithm 1 Modified Adaptive Heterogeneous Comprehensive Particle Swarm Optimization (mAHPSO)

Notations:

- N: the population size
- N_1, N_2: subpopulation sizes
- c_1, c_2: acceleration coefficients
- w: inertia weight
- $w_{\text{min}}, w_{\text{max}}$: minimum and maximum inertia weights
- T_{max}: maximum number of iterations
- D: dimensionality
- FEs: the number of fitness evaluations
- MaxFEs: the maximum number of fitness evaluations allowed
- X_i: particle's position
- V_i: particle's velocity
- $f(X_i)$: fitness value of particle X_i
- P_{adjust}: adjustment probability
- g_{best}: the best solution found so far
- $g_{\text{best\_prev}}$: the previous best solution
- $flag$: a flag variable to track certain conditions
- AIW: Adaptive Inertia Weight strategy
- DOL: Diversity-Enhancing Oppositional Learning strategy
- ADM: Adaptive Dynamic Mutation strategy

Initialization: Set the parameters: N, N_1, N_2, acceleration $c_1, c_2, w, w_{\text{min}}, w_{\text{max}}, T_{\text{max}}$, and D.
Calculate X_i, V_i, and $f(X_i)$.
while $FEs \leq$ MaxFEs **do**
  for $i = 1$ **to** N_1 **do**
    Calculate the average fitness M of the subpopulation.
    Calculate the inertia weight w.
    Update the exploitation subpopulation X and V.
  end for
  for $i = N_1 + 1$ **to** N **do**
    Update the exploration subpopulation X and V.
    Use DOL strategy.
    Calculate the adjustment probability P_{adjust}.
    if rand $\geq P_{\text{adjust}}$ **then**
      Execute ADM strategy [23] .
    else
      Execute ADM strategy [23].
    end if
    Update g_{best} [23].
    $FEs \leftarrow FEs + 1$
  end for
  Update $g_{\text{best\_prev}}$ and flag variable using appropriate Eqs. (1) and (2) respectively.
  if flag == max-iter **then**
    return
  end if
end while

adjust its search strategy, improving its ability to find optimal solutions efficiently and effectively. By balancing exploration and exploitation more adeptly, dynamic phase switching makes mLSHADE a powerful tool for solving a wide range of optimization problems.

LSHADE also enhances the SHADE algorithm by continuously reducing the population size linearly throughout the search process [15]. SHADE is a variant of Differential Evolution (DE) that adapts its control parameters based on historical memory. LSHADE extends this by dynamically resizing the population, ensuring that the population size decreases from its initial size to a minimum size over the course of the optimization process. This reduction helps balance exploration and exploitation, improving convergence and preventing the algorithm from getting stuck in local optima.

The LSHADE algorithm builds on the Success-History Based Adaptive Differential Evolution (SHADE) by incorporating a Linear Population Size Reduction (LPSR) mechanism. SHADE uses historical memory to adapt its control parameters, such as the crossover rate and scaling factor, based on past successful values. LSHADE extends this by gradually reducing the population size in a linear manner throughout the optimization process. This reduction starts from an initial population size and decreases to a predefined minimum size, balancing exploration and exploitation more effectively. By continuously adjusting the population size, LSHADE enhances the algorithm's ability to converge to optimal solutions without getting trapped in local optima.

Algorithm (2) represents the complete mLSHADE algorithm with all its phases, including initialization, main loop with population updates, trial vector evaluation, archive management, and memory updates. The highlighted part of the algorithm represents the incorporation of dynamic phase switching, which ensures the algorithm avoids premature convergence and enhances its robustness in finding optimal solutions.

3.4 Hybrid L-SHADE and AHPSO Using Adaptive Switching Mechanism (hL-AHSPO-SM)

The proposed hybrid algorithm is divided into five stages.

Initialization

Initially, all the particles in the population are randomly positioned using continuous uniform distribution, where each particle represents a candidate solution. The equation for each individual's position $X_{i,j}$ in the matrix *X*, is defined in Eq. (4):

$$X_{i,j} = x_{\min} + \text{rand}_{i,j} \times (x_{\max} - x_{\min}) \tag{4}$$

Algorithm 2 Modified LSHADE Algorithm (mLSHADE)

Notations:

- G: The current generation.
- N_G: Population size at generation G.
- A: Archive set.
- P_G: Population at generation G.
- M_{CR}, M_F: Memory arrays for crossover rate and scaling factor, respectively.
- S_{CR}, S_F: Successful crossover rate and scaling factor sets.
- r_i: Index selected from [1, H] randomly.
- randn$_i(M_{CR,r_i}, 0.1)$: Normally distributed random number with mean M_{CR,r_i} and standard deviation 0.1.
- randc$_i(M_{F,r_i}, 0.1)$: Cauchy distributed random number with location parameter M_{F,r_i} and scale parameter 0.1.
- $u_{i,G}$: Trial vector.
- $x_{i,G}$: Parent vector.
- $f(x)$: Objective function value of vector x.

```
Initialization: G = 1, NG = N_init, Archive A = ∅, MCR = MF = 0.5, PG = ...
while FEs ≤ MaxFEs do                                              ▷ Main loop
    SCR = ∅, SF = ∅;
    for i = 1 : N do                                               ▷ Loop over population
        ri = Select randomly [1, H] randomly;
        if MCR, ri = ⊥ then
            CR_{i,G} = 0;
        else
            CR_{i,G} = randni(MCR, ri, 0.1);
        end if
        Fi, G = randci(MF, ri, 0.1);
    end for
    for i = 1 : N do                                               ▷ Loop for trial vector evaluation
        if f(u_{i,G}) ≤ f(x_{i,G}) then
            x_{i,G} − u_{i,G};
        else
            x_{i,G+1} = x_{i,G};
        end if
        if f(u_{i,G}) < f(x_{i,G}) then
            x_{i,G} → A;
            CR_{i,G} → SCR, Fi, G → SF,
        end if
    end for
    If necessary, delete randomly selected individuals from the archive such that the archive size is |A|.
    Update memories MCR and MF [15];
    Calculate NG + 1;
    if NG < NG + 1 then
        Sort individuals in P based on their fitness values and delete lowest NG − NG + 1 members;
        Resize archive size |A| according to new |P|;
    end if
    Update the gbest_prev and flag variable using Eq. (1) and Eq. (2) respectively.
    if flag == max-iter then
        return
    end if
end while
```

where

- $x_{\max}$, $x_{\min}$: represents upper and lower bound positions of the individuals
- $\text{rand}_{\text{i,j}}$: represents uniformly distributed random number between 0 and 1 for the i-th particle and j-th dimension.

The equation for each individual's velocity $V_{i,j}$ in the matrix V is defined in Eq. (5):

$$V_{i,j} = v_{\min,j} + \text{rand}_{i,j} \times (v_{\max,j} - v_{\min,j}) \tag{5}$$

where

- $x_{\max}$, $x_{\max}$: represents the maximum and minimum velocity of the individuals
- $rand_{\text{i,j}}$: represents uniformly distributed random number between 0 and 1 for the i-th particle and j-th dimension.

After the calculation of each individual's position and velocity the initial fitness of the population is evaluated. In this process the personal best of each individual, pBest is stored in another matrix and the global best, gBest is also updated, to set as a benchmark for the iterations that follow. The pbest for each individual and the gBest is updated in every iterations that follows.

mAHPSO Phase

Initialized particles are provided to the mAHPSO algrithm during the initial phase. The mAHPSO effectively maintains a balance between exploration and exploitation. The search space is traversed iteratively to identify optimal solutions. This equilibrium ensures that the algorithm does not unduly converge to unsatisfactory solutions and that it continues to explore new areas of the search space. Upon detecting stagnation, the algorithm implements a dynamic phase-switching mechanism to exchange strategies, thereby facilitating the identification of innovative solutions. The flexibility of mAHPSO enables it to effectively address a wide range of optimization challenges.

First Intermediate Phase

Once AHPSO detects a phase change, the algorithm switches to an intermediate phase. This stage is crucial for establishing the mLSHADE algorithm. The algorithm chooses a dynamic subset of individuals with varying size, consisting of the top-performing solutions from the overall population. The L-SHADE algorithm employs Linear Population Size Reduction (LPSR) to dynamically alter the population size between generations. This modification centers on systematically diminishing the population size and directing the search towards favorable places within the search space. The size of the transitioning subpopulation is determined by several factors, including individual performance (which favors individuals closer to optimal solutions), problem dimension (higher dimensions require a larger subpopulation for sufficient exploration), and population performance variability (higher variability requires a larger subpopulation to prevent the overlooking of promising regions).

Considering these factors, the function to calculate the length of the subpopulation, L, is defined in Eq. (6):

$$L = \max(L_{\min}, \min(L_{\max}, \lceil P \cdot f(D, V) \rceil)) \tag{6}$$

where:

- L is the resulting value.
- $L_{\min}$ and $L_{\max}$ are the minimum and maximum limits for allowed sizes of subpopulation respectively.
- P is total population size.
- $f(D, V)$ is a function of the problem's dimensionality (D) and the population's performance variability (V).
- $\lceil \cdot \rceil$ denotes the ceiling function, which rounds up to the nearest integer.

The function $f(D, V)$ is defined in Eq. (7):

$$f(D, V) = \alpha \cdot \frac{D_{\max}}{D} + \beta \cdot \frac{V_{\max}}{V} \tag{7}$$

where:

- α and β are weighting factors for dimensionality and variability, respectively.
- $D_{\max}$ and $D_{\max}$ are normalization factors.

According to the subpopulation size function, we create the most optimal individual by arranging the entire population in order and choosing the top individuals for the next algorithm which will be provided to the mLSHADE algorithm in the next phase.

mLSHADE Phase

During this stage, the reduced population is transferred to the mLSHADE algorithm. mLSHADE employs an iterative approach to explore the search space in order to find the best solutions, while simultaneously reducing the population size to minimize the number of function evaluations. In the event of stagnation, the algorithm employs the dynamic phase-switching mechanism to investigate novel options, so ensuring uninterrupted advancement towards optimal solutions. The mLSHADE algorithm also returns the archive memory in order in regenerate the reduced population.

Second Intermediate Phase

After mLSHADE phase, the algorithm transitions to another intermediate phase. The population returned by mLSHADE cannot be directly used due to L-SHADE's LPSR strategy, which results in a smaller population size. For this we calculate the diference between the initial subpopulation size and the population returned from mLSHADE algorithm. The result provides the number of individuals required to regain the original population size. The algorithm employs one of two strategies to regain the reduced population by L-Shade algorithm based on its current phase:

- Gradient Repair: If the algorithm is in its first half, it uses gradient repair to restore the reduced population using the L-SHADE's archive memory. The gradientRepair method ensures that the population size is maintained by either selecting a subset of the existing archive memory returned from the mLSHADE algorithm or generating new random solutions if the archive memory size is less than the archive memory size. This approach helps in restoring the population size to its required level, thus maintaining the algorithm's robustness and ensuring it can continue iterating effectively.
- gBest Mutation: If the algorithm is in its second half, it uses gBest mutation to repopulate the reduced population. The gBest Mutation method generates new candidate solutions by applying random perturbations to the global best solution found so far. This approach helps in exploring new regions of the solution space while leveraging the high-quality solution provided by gbest. The mutation strength sigma controls how much variation is introduced, allowing the algorithm to balance exploration and exploitation effectively. In mathematical terms, if `gBest` is the best solution vector, as shown in Eq. (7):

$$\mathbf{g} = [g_1, g_2, \ldots, g_d] \tag{7}$$

where d is the number of dimensions. The new solutions can be represented as shown in Eq. (8):

$$\mathbf{x}_i = \mathbf{g} + \sigma \cdot \mathbf{z}_i \tag{8}$$

for $i = 1, 2, \ldots, \mathbf{N_R}$, where $\mathbf{z}_i$ is a vector of random numbers drawn from a standard normal distribution, and σ is the mutation strength.
This choice ensures the algorithm adapts its strategy based on its progress, maintaining robustness and efficiency.

The modified L-SHADE population is then recombined with the best individuals selected for the L-SHADE phase, and the algorithm returns back to AHPSO Phase, ready to iterate further.

Figure 1 shows the process work flow of the proposed algorithm. By structuring the algorithm into these detailed phases, it ensures a balanced, adaptive, and efficient search for optimal solutions, leveraging both AHPSO's exploratory capabilities and L-SHADE's exploitative focus.

The overall algorithm of Hybrid L-SHADE and Adaptive heterogeneous comprehensive learning particle swarm optimization with switching mechanism (HL-AHPSO-SM) is in Algorithm (3).

4 Experiments and Results

4.1 Benchmark Functions

There are ten test functions $f_1 - f_{10}$ with four different dimensions: $D = 5, 10, 15$, and 20 in the CEC2020 Single Objective Bound Constrained Optimization Competition [8]. Table 1 describes the overview of all the functions used for the Evaluation of the proposed algorithm. The benchmark function consists of 2 unimodal functions, 3 basic functions, 2 hybrid functions and 3 composition functions. The search ranges of decision variables are limited to $[-100, 100]$ for all functions.

4.2 Parameter Setting

The hL-AHPSO-SW algorithm combines L-SHADE and AHPSO with a switching mechanism to optimize benchmark functions. The population size is dynamically adjusted based on the dimensionality of the problem, with a minimum size of 40. The algorithm uses specific parameters for L-SHADE and AHPSO, including a success history memory size of 6, and velocity and inertia weight ranges for AHPSO. This hybrid approach aims to leverage the strengths of both L-SHADE and AHPSO to achieve optimal performance on the CEC 2020 benchmark problems (Table 2).

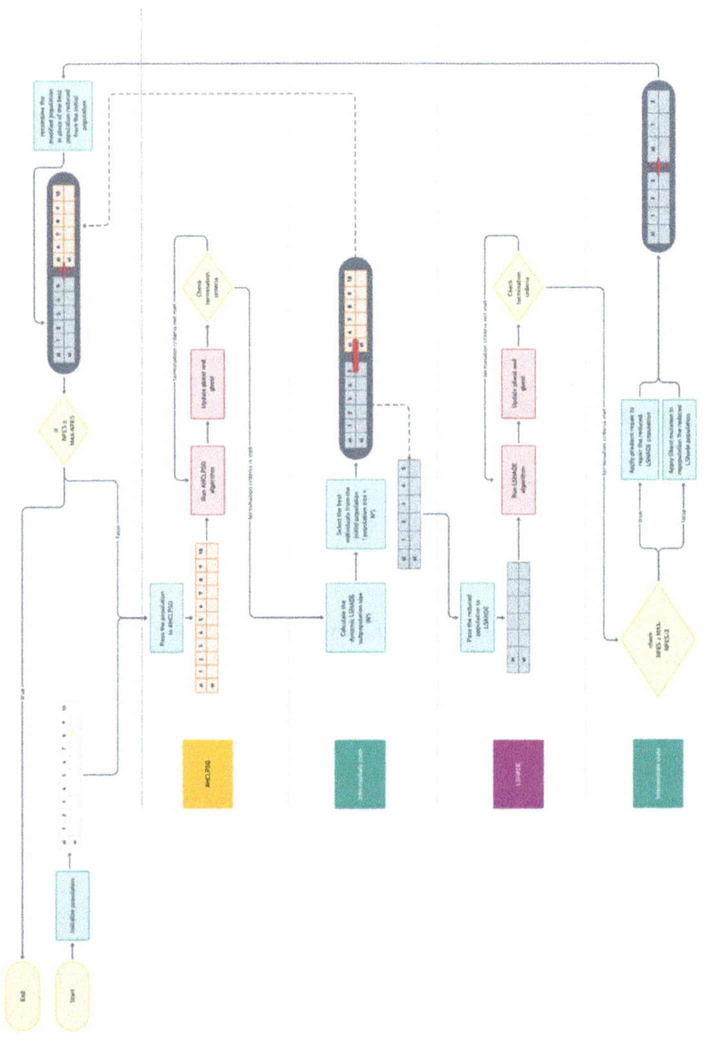

Fig. 1 The process flow of the proposed hL-AHPSO-SM

Table 1 Overview of functions used in the algorithm evaluation

Function type	No	Functions	$Fi = Fi(x*)$
	1	Shifted and rotated bent cigar function	100
Unimodal function	2	Shifted and rotated Schwefels function	1100
	3	Shifted and rotated Lunacek bi-Rastrigin function	700
	4	Expanded Rosenbrocks plus Griewangks function	1900
Basic functions	5	Hybrid function 1 ($N = 3$)	1700
	6	Hybrid function 2 ($N = 4$)	1600
Hybrid functions	7	Hybrid function 3 ($N = 5$)	2100
	8	Composition function 1 ($N = 3$)	2200
	9	Composition function 2 ($N = 4$)	2400
Composition function	10	Composition function 3 ($N = 5$)	2500

Table 2 Value of parameters in the proposed algorithm

Parameter	Meaning	Value
N_{init}	Size of the initial population	$20D$
N_{min}	Minimal sub-population size	40
H	Size of the success history memory	6
r_{arc}	Archive size $\lceil 4 \cdot p \cdot \text{round}(r_{\text{arc}} \cdot N_{\text{pop}}) \rceil$	2.6
p	Required in cur-to-pbest/1 mutation	0.11
lu	Maximum number of iterations for DPS	20
$T_{\max}$	Lower and upper bounds for position	[−100,100]
Co	Parameter for mutation control in L-SHADE	0.5
Fo	Parameter for learning factor in L-SHADE	0.5
$v_{\min}$	Minimum velocity	−1
$v_{\max}$	Maximum velocity	1
$w_{\min}$	Maximum inertia weight	0.9
$w_{\max}$	Minimum inertia weight	0.2
c	Constant for adjusting inertia weight	0.008
$t_{\text{divisions}}$	Number of time divisions for dynamic inertia weight	10

Algorithm 3 HL-AHPSO-SM Algorithm

Notations:

- D: dimension
- MaxFEs: the maximum number of fitness evaluations allowed
- N: the population size
- $x_{\min}$, $x_{\max}$: minimum and maximum
- c: constant for adjusting interia weight
- w: inertia weight
- $v_{\min}$, $v_{\max}$: minimum and maximum velocities
- $w_{\min}$, $w_{\max}$: minimum and maximum inertia weights
- $T_{\max}$: maximum number of iterations for dynamic phase switching
- FEs: the number of fitness evaluations
- X_i: particle's position
- V_i: particle's velocity
- g_{best}: the best solution found so far

Initialization: D, MaxFEs, N, $x_{\min}$, $x_{\max}$, $v_{\min}$, $v_{\max}$, $w_{\min}$, $w_{\max}$, c
Generate initial population X and velocities V for AHPSO.
while FEs $\leq$ MaxFEs **do** ▷ Main loop
    if phase == AHPSO **then**
        Call mAHPSO function to perform AHPSO with switching mechanism
        Update gbest, gbest_val, FEs, X, pbest, phase
        Sort pbest based on fitness values
    else if phase == LSHADE **then**
        Calculate the dynamic subpopulation size using eq.(5).
        Select best performing subpopulation from AHPSO
        Call mLSHADE function to perform L-SHADE with switching mechanism
        Update gbest, gbest_val, FEs, X, phase
        if FEs > MaxFEs/2 **then**
            Generate new solutions using gbest mutation
        else
            Repair discarded memory using gradient repair
        end if
        Update population X with new solutions
    end if
end while

4.3 Algorithm Complexity

Algorithm complexity is estimated by calculating T_0, T_1, and T_2 values, as defined in the CEC2020 competition [8]. It aims to show the required computational effort of an algorithm with respect to different problem dimensions. T_0 denotes the time required to run the following computation for 1,000,000 times.

Table 3 Computational complexity

Problem Dimension	T_0 (ms)	T_1 (ms)	T_2 (ms)	$(T_2 - T_1)/T_0$
$D = 5$	0.514	51.6105	219.9182	210.25
$D = 10$	0.8005	70.9715	262.6592	239.46
$D = 15$	1.101	101.8337	337.0386	293.82

$$x = x + x/2; \quad x = x * x; \quad x = \sqrt{x};$$
$$x = \log(x); \quad x = \exp(x); \quad x = x/(x+2)$$

T_1 denotes the time required to evaluate function f_1 for 200,000 times, and T_2 denotes the average computation time (over 5 runs) to run a whole evolutionary algorithm with MAX_NFE = 200,000. The competition rules ask for the measured values for $D = 5, 10,$ and 15, shown in Table 3.

The experiments were performed in a Windows 10 environment with Ryzen 7 4800H (2.90GHz) CPU and 16GB DDR4-3200 RAM. The hL-AHPSO-SM was implemented in MATLAB.

Asymptotic Complexity

In addition to the wall-clock measurements T_0–T_2, we present asymptotic (Big-O) estimates for the main components of the algorithms to clarify their computational scaling.

In each iteration, AHPSO updates the positions and velocities of all N particles across D dimensions, evaluates fitness, and applies subpopulation-specific strategies such as Adaptive Inertia Weight (AIW), Adaptive Dimension Mutation (ADM), and Diversity-Enhancing Oppositional Learning (DOL). Thus, a single iteration costs $O(N \cdot D)$, and for T iterations the asymptotic cost is:

$$T_{\text{AHPSO}} = O(N \cdot D \cdot T). \tag{9}$$

L-SHADE (a DE variant) generates and evaluates trial vectors, updates memory arrays, and manages the archive at each generation. For population size N, dimension D, and G generations:

$$T_{\text{L-SHADE}} = O(N \cdot D \cdot G). \tag{10}$$

The proposed hybrid alternates between AHPSO and L-SHADE, transferring a subpopulation of size L between them. If T_A and T_L are the respective iteration counts, the total complexity is:

Table 4 Asymptotic (Big-O) complexity of algorithms

Algorithm	Asymptotic time complexity
AHPSO (per run)	$O(N \cdot D \cdot T)$
L-SHADE (per run)	$O(N \cdot D \cdot G)$
hL-AHPSO-SM (Hybrid)	$O(N \cdot D \cdot T_A + L \cdot D \cdot T_L + C_{\text{switch}})$

Table 5 The maximum number of fitness function evaluations

Problem dimension	*MAX_NFE*
$D = 5$	50,000
$D = 10$	1,000,000
$D = 15$	2,000,000
$D = 20$	10,000,000

$$T_{\text{Hybrid}} = O\big((N \cdot D \cdot T_A) + (L \cdot D \cdot T_L) + C_{\text{switch}}\big), \tag{11}$$

where C_{switch} represents the overhead of dynamic phase switching, sorting, and archive repair, which is linear in $N \cdot D$ per switch. Since $L \leq N$, the hybrid s asymptotic behavior remains similar to its constituent algorithms, with only a small linear overhead (Table 4).

The asymptotic expressions describe the scaling behavior of each algorithm, whereas the empirical values in Table 3 reflect real execution times on the given hardware and implementation (MATLAB). Both analyses together provide a balanced view: asymptotic complexity shows theoretical growth trends, while measured timing demonstrates the practical computational effort for given dimensions.

4.4 Numerical Results

The competition defines the maximum number of fitness function evaluations (MAX_NFE) for the four problem dimensions, listed in Table 3. The algorithm solved each test function for 30 times. According to the competition rules, statistics including the best, the worst, median, mean, and standard deviation of the error values over 30 runs are reported in Tables 5, 6, 7 and 8. The error value is the deviation of the obtained solution from the optimum. If the error value is less than 10^{-8}, it is regarded as zero.

When the problem dimension is small ($D = 5$), our algorithm solves functions f_1 and f_5 to the optimality all the time. The optimal solutions for functions f_2, f_3, f_6, and f_7 can be found in the best case. When the dimension increases to 10 or higher, function f_1 still can be solved optimally over all 30 runs; functions f_5, f_8, and f_9 are solvable when $D = 10$, and f_3 is solvable when $D = 15$ and 20. Functions f_2,

Table 6 Error values of HL-AHPSO-SM ($D = 5$)

Func.	Best	Worst	Median	Mean	Std
f_1	0.0000E+00	0.0000E+00	0.0000E+00	0.0000E+00	0.0000E+00
f_2	0.0000E+00	5.1230E−01	8.7600E−02	1.3670E−01	1.2876E+00
f_3	0.0000E+00	2.0000E−05	1.2000E−03	9.0000E−04	1.5344E+00
f_4	3.2100E−02	1.4980E−01	1.0920E−01	9.6300E−02	3.1200E−02
f_5	0.0000E+00	0.0000E+00	0.0000E+00	0.0000E+00	0.0000E+00
f_6	0.0000E+00	1.1321E+02	9.8760E+01	3.7820E+01	1.9450E+01
f_7	0.0000E+00	1.0543E+02	9.9870E+01	1.0371E+02	4.1230E+01
f_8	0.0000E+00	3.5687E+02	3.5298E+02	3.4129E+02	1.6930E+01
f_9	1.0000E−04	0.0000E+00	1.0000E−04	1.0000E−04	1.0000E−04
f_{10}	0.0000E+00	0.0000E+00	0.0000E+00	0.0000E+00	0.0000E+00

Table 7 Error values of HL-AHPSO-SM ($D = 10$)

Func.	Best	Worst	Median	Mean	Std
f_1	0.0000E+00	0.0000E+00	0.0000E+00	0.0000E+00	0.0000E+00
f_2	7.0123E−02	3.7123E+00	2.6291E−01	6.9812E−01	1.2123E+00
f_3	1.1321E−01	1.1321E+01	1.0231E+01	1.0567E+01	2.2345E+01
f_4	1.7865E−01	4.1321E−01	2.8765E−01	2.9421E−01	6.0123E−02
f_5	0.0000E+00	1.5321E−01	2.2091E−01	1.3231E−01	1.1891E−01
f_6	1.5732E+01	1.3321E+01	4.0091E−02	5.1234E−02	4.1091E−02
f_7	8.1234E−06	4.4321E−01	2.8091E−02	1.3891E−01	1.5651E−01
f_8	0.0000E+00	1.0132E+02	1.0321E+02	9.7891E+01	1.8321E+01
f_9	0.0000E+00	1.4121E+02	3.7321E+02	4.3212E+01	9.3121E+01
f_{10}	9.8234E+02	4.9321E+02	4.1321E+02	4.1891E+02	1.4321E+01

Table 8 Error values of HL-AHPSO-SM ($D = 15$)

Func.	Best	Worst	Median	Mean	Std
f_1	0.0000E+00	0.0000E+00	0.0000E+00	0.0000E+00	0.0000E+00
f_2	0.0000E+00	2.3610E+00	1.2451E+01	1.8276E+01	4.1725E−01
f_3	1.5627E−01	1.5567E−01	1.5547E−01	1.5567E−01	2.0937E−13
f_4	2.4414E−01	4.5490E−01	3.9589E−01	3.8242E−01	4.4318E−02
f_5	1.5622E−01	2.1480E−01	3.1272E−01	5.0472E−01	5.1472E−01
f_6	4.6565E−01	9.6658E−01	7.3095E−01	7.2717E−01	1.2695E−01
f_7	1.7399E−01	9.0284E−01	5.0020E+01	4.5520E−01	1.0598E−01
f_8	1.0020E+02	1.0020E+02	1.0020E+02	1.0020E+02	5.0118E−13
f_9	3.7927E+02	3.8926E+02	3.8965E+02	4.7694E−02	1.1598E−01
f_{10}	4.0020E+02	4.0020E+02	4.0020E+02	4.0020E+02	5.0815E−13

Table 9 Error values of HL-AHPSO-SM ($D = 20$)

Func.	Best	Worst	Median	Mean	Std
f_1	0.0000E+00	0.0000E+00	0.0000E+00	0.0000E+00	0.0000E+00
f_2	0.0000E+00	9.3685E−02	3.1282E−02	3.7632E−02	1.1195E−02
f_3	2.0387E+01	2.0387E+01	2.0387E+01	2.0387E+01	4.7636E−13
f_4	4.1611E−01	6.5582E−01	5.0096E−01	4.9841E−01	1.3423E−02
f_5	2.0817E−01	6.8361E+00	1.2525E+00	1.3081E+00	1.4521E+02
f_6	7.9202E−02	2.9246E−01	2.0161E−01	2.5022E−01	4.7101E−02
f_7	2.5517E−02	1.5000E+01	4.8157E−02	9.9948E−01	1.1784E−01
f_8	4.0000E+02	4.0000E+02	4.0000E+02	4.0000E+02	4.9831E−13
f_9	4.0000E+02	4.0000E+02	4.0000E+02	4.0000E+02	4.9831E−13
f_{10}	4.1366E+02	4.1366E+02	4.1366E+02	4.1366E+02	7.4135E−04

f_6, f_7, and f_{10} are difficult to our algorithm. It almost gets trapped in the same local optimum all the time.

4.5 Performance Comparison

To know how well the proposed HL-AHPSO-SM performs, we also executed L-SHADE and mlL-SHADE to solve each of the ten test functions 60 times with the computational budget in Table 4. In Tables 5, 6 and 7, we list the mean and standard deviation of error values for the three algorithms. The standard deviation values are listed in parentheses. We performed a t-test at the significance level of 0.05. The symbols +, −, and ≈ indicate that the proposed HL-AHPSO-SM is better than (+), worse than (−), and not statistically different from (≈) the compared algorithm (Tables 8, 9, 10, 11 and 12).

D	mpml_SHADE vs.			≈	=
	ml_SHADE	L-SHADE	ml_SHADE ≈ L-SHADE		
5	1	0	0	0	0
10	5	5	7	3	0
15	4	1	6	4	0
20	3	6	7	2	1
Total	14	22	20	17	1

Table 10 Mean and std of error values ($D = 5$)

Func.	HL-AHPSO-SM	mlL-SHADE	L-SHADE
f_1	0 (0)	0 (0) ≈	0 (0) ≈
f_2	0.1367 (1.2876)	0.2349 (0.8698) +	0.3109 (0.8785) +
f_3	0.0009 (1.5344)	4.9887 (1.0875) +	4.9091 (1.4654) +
f_4	0.0963 (0.0312)	0.0872 (0.0405) +	0.0976 (0.0380) ≈
f_5	0 (0)	0 (0) ≈	0.0104 (0.0799) +
f_6	37.820 (19.4500)	0 (0) +	1.667 (12.8019) –
f_7	103.71 (41.2300)	101.6744 (12.8610) –	101.6667 (12.8019) –
f_8	341.29 (16.9300)	344.2091 (11.8145) +	341.8408 (15.2059) ≈
f_9	0.0001 (0.0001)	0 (0) ≈	0 (0) ≈
f_{10}	0 (0)	0 (0) ≈	0 (0) ≈

Table 11 Mean and std of error values ($D = 10$)

Func.	HL-AHPSO-SM	mlL-SHADE	L-SHADE
f_1	0 (0)	0 (0)	0 (0) ≈
f_2	0.69812 (1.2123)	1.9702 (2.4104) +	5.5230 (3.9118) +
f_3	10.567 (22.345)	10.7776 (0.5729) ≈	11.8947 (0.5659) +
f_4	0.29421 (0.060123)	0.2909 (0.0497) ≈	0.3000 (0.0567) ≈
f_5	0.13231 (0.11891)	0.3763 (0.3927) +	0.2492 (0.1835) +
f_6	0.051234 (0.041091)	0.1047 (0.1541) ≈	0.2395 (0.1640) +
f_7	0.13891 (0.15651)	0.0465 (0.1207) ≈	0.2531 (0.2346) +
f_8	97.891 (18.321)	98.7399 (9.6793) ≈	98.3391 (12.8027) +
f_9	43.212 (93.121)	284.9399 (87.9440) +	296.4128 (82.5192) +
f_{10}	418.91 (14.321)	412.5086 (21.3484) –	413.2784 (21.6126)

Table 12 Mean and std of error values ($D = 15$)

Func.	HL-AHPSO-SM	mlL-SHADE	L-SHADE
f_1	0 (0)	0 (0) ≈	0 (0) ≈
f_2	18.276 (0.41725)	0.3952 (0.7628) –	8.6542 (16.1597) –
f_3	0.15567 (2.0937e–13)	15.7586 (0.3642) +	16.7030 (0.4934) +
f_4	0.38242 (0.044318)	0.3965 (0.0438) ≈	0.3734 (0.0419) ≈
f_5	0.50472 (0.51472)	1.1141 (1.0080) +	7.4813 (25.8315) +
f_6	0.72717 (0.12695)	0.8408 (0.1842) +	1.2232 (1.1271) +
f_7	0.45520 (0.10598)	0.6679 (0.1887) +	0.7104 (0.2051) +
f_8	100.2 (5.0118e–13)	100 (9.6793) ≈	100 (0) ≈
f_9	0.047694 (0.11598)	382.6355 (23.0764) +	390.0513 (0.2892) +
f_{10}	400.2 (5.0815e–13)	400 (0) ≈	400 (0) ≈

5 Conclusion and Future Work

This chapter proposed a hybrid metaheuristic named hL-AHPSO-SM, which integrates Adaptive Heterogeneous Comprehensive Learning PSO (AHPSO) and L-SHADE through a dynamic switching mechanism. The hybridization strategy was designed to balance exploration and exploitation by adaptively transferring individuals between the PSO-based and DE-based phases. The switching mechanism allows the algorithm to dynamically adjust its search behavior depending on the population s progress, thereby preventing premature convergence and maintaining population diversity.

Experimental evaluation on the CEC2020 single-objective benchmark functions demonstrated that the proposed hL-AHPSO-SM achieves competitive and often superior performance compared to the baseline algorithms L-SHADE and mlL-SHADE. The results show that the hybrid s adaptive switching mechanism improves convergence stability and overall solution quality across diverse problem landscapes.

In the present work, comparative performance was discussed primarily using mean and standard deviation of error metrics. Although these statistics provide a general indication of algorithmic behavior, statistical hypothesis testing would offer a more objective validation of the observed differences. As part of future work, we plan to incorporate non-parametric statistical tests such as the Wilcoxon signed-rank test and Friedman test to substantiate the comparative findings. This addition will help quantify the significance of performance improvements achieved by the proposed hL-AHPSO-SM and enhance the interpretability of its outcomes across benchmark functions.

Further research will also focus on extending the proposed hybrid strategy to multi-objective and constrained optimization problems, integrating adaptive control of switching frequency, and exploring the use of surrogate models to reduce computational cost in large-scale problems.

Data Availability Statement The data that support the findings of this study are openly available in https://github.com/Birajj2189/HL-AHPSO-SM.

References

1. Bianchi, L., Dorigo, M., Gambardella, L.M., Gutjahr, W.J.: A survey on metaheuristics for stochastic combinatorial optimization. Nat. Comput. **8**, 239–287 (2009)
2. Brest, J., Maučec, M.S., Bošković, B.: il-shade: improved l-shade algorithm for single objective real-parameter optimization. In: 2016 IEEE Congress on Evolutionary Computation (CEC), pp 1188–1195. IEEE (2016)
3. Črepinšek, M., Liu, S.-H., Mernik, M.: Exploration and exploitation in evolutionary algorithms: a survey. ACM Comput. Surv. (CSUR) **45**(3), 1–33 (2013)
4. Eiben, A.E., Schippers, C.A.: On evolutionary exploration and exploitation. Fund. Inform. **35**(1–4), 35–50 (1998)
5. Fausto, F., Reyna-Orta, A., Cuevas, E., Andrade, Á.G., Perez-Cisneros, M.: Metaheuristics for all tastes: from ants to whales. Artif. Intell. Rev. **53**, 753–810 (2020)

6. Hiermann, G., Prandtstetter, M., Rendl, A., Puchinger, J., Raidl, G.R.: Metaheuristics for solving a multimodal home-healthcare scheduling problem. CEJOR **23**, 89–113 (2015)
7. Huang, C., Jinhao, Yu., Yang, J.: Top-level dual exploitation particle swarm optimization. Memetic Comput. **16**(1), 1–28 (2024)
8. Jou, Y.-C., Wang, S.-Y., Yeh, J.-F., Chiang, T.-C.: Multi-population modified l-shade for single objective bound constrained optimization. In: 2020 IEEE Congress on Evolutionary Computation (CEC), pp 1–8. IEEE (2020)
9. Kennedy, J., Eberhart, R.: Particle swarm optimization. In: Proceedings of ICNN'95-international conference on neural networks, vol 4, pp 1942–1948. IEEE (1995)
10. Lynn, N., Suganthan, P.N.: Heterogeneous comprehensive learning particle swarm optimization with enhanced exploration and exploitation. Swarm Evol. Comput. **24**, 11–24 (2015)
11. Pant, M., Zaheer, H., Garcia-Hernandez, L., Abraham, A., et al.: Differential evolution: A review of more than two decades of research. Eng. Appl. Artif. Intell. **90**, 103479 (2020)
12. Storn, R., Price, K.: Differential evolution-a simple and efficient heuristic for global optimization over continuous spaces. J. Global Optim. **11**, 341–359 (1997)
13. Suganthan, P.N.: Particle swarm optimiser with neighbourhood operator. In: Proceedings of the 1999 congress on evolutionary computation-CEC99 (Cat. No. 99TH8406), vol. 3, pp. 1958–1962. IEEE (1999)
14. Tanabe, R., Fukunaga, A.: Success-history based parameter adaptation for differential evolution. In: Proceedings of IEEE congress on evolutionary computation (CEC), pp 71–78 (2013)
15. Tanabe, R., Fukunaga.: Improving the search performance of shade using linear population size reduction. In: 2014 IEEE congress on evolutionary computation (CEC), pp 1658–1665. IEEE (2014)
16. Tang, H., Lee, J.: Adaptive initialization lshade algorithm enhanced with gradient-based repair for real-world constrained optimization. Knowl.-Based Syst. **246**, 108696 (2022)
17. Tsai, C.-W., Chiang, M.-C., Ksentini, A., Chen, M.: Metaheuristic algorithms for healthcare: open issues and challenges. Comput. Electr. Eng. **53**, 421–434 (2016)
18. Verma, P., Parouha, R.P.: An advanced hybrid meta-heuristic algorithm for solving small-and large-scale engineering design optimization problems. J. Electr. Syst. Inform. Technol. **8**, 1–43 (2021)
19. Xiong, N., Molina, D., Ortiz, M.L., Herrera, F.: A walk into metaheuristics for engineering optimization: principles, methods and recent trends. Int. J. Comput. Intell. Syst. **8**(4), 606–636 (2015)
20. Xu, J., Zhang, J.: Exploration-exploitation tradeoffs in metaheuristics: Survey and analysis. In: Proceedings of the 33rd Chinese control conference, pp. 8633–8638. IEEE (2014)
21. Yang, X.-S.: Optimization and metaheuristic algorithms in engineering. Metaheurist. Water Geotech. Transp. Eng. **1**, 23 (2013)
22. Yang, X., Li, H.: Multi-sample learning particle swarm optimization with adaptive crossover operation. Math. Comput. Simul. **208**, 246–282 (2023)
23. Yang, X., Li, H., Xia, Yu.: Adaptive heterogeneous comprehensive learning particle swarm optimization with history information and dimensional mutation. Multimed. Tools Appl. **82**(7), 9785–9817 (2023)
24. Yeh, J.-F., Chen, T.-Y., Chiang, T.-C.: Modified l-shade for single objective real-parameter optimization. In: 2019 IEEE congress on evolutionary computation (CEC), pp. 381–386. IEEE (2019)

Evolutionary Multi and Many-Objective Optimization: Enhancements Using Machine Learning

Dhish Kumar Saxena, Sukrit Mittal, and Kalyanmoy Deb

Abstract Evolutionary Algorithms (EAs) are known to efficiently approximate the Pareto-front (PF) in a single simulation run for problems with conflicting objectives. Conventionally, problems with two- or three-objectives are referred to as multi-objective problems, while those with four or more objectives are referred to as many-objective problems. EAs capable of handling both these problem classes are referred to as Evolutionary Multi- and Many-objective Optimization Algorithms (EMâOAs). Despite their success, researchers continually challenge EMâOAs with increasingly complex real-world problem characteristics, where ensuring convergence to and diversity across the Pareto-optimal front (PF) becomes difficult. Most efforts to improve EMâOAs' efficiency have focused on improving the inherent bio-inspired operators. However, treatment of EMâOAs' generational populations as data sets amenable to Machine Learning (ML) towards alleviating converge and diversity challenges is an emergent trend. As proponents of the latter, the authors' research has demonstrated that ML techniques can utilize the generational non-dominated solutions to (i) *discern* the underlying problem structure, potentially facilitating dimensionality reduction, (ii) *learn* efficient search trajectories to expedite convergence towards Pareto-optimal (PO) solutions, (iii) *learn* mechanisms to improve diversity among PO solutions, and (iv) *generate* new non-dominated solutions from the final PO solutions to help find the preferred solution by the decision-maker. This chapter, while introducing a holistic perspective, dominantly focuses on presenting how ML techniques can enhance the search efficiency of EMâOAs in terms of convergence

D. K. Saxena (✉)
Indian Institute of Technology Roorkee, Roorkee, India
e-mail: dhish.saxena@me.iitr.ac.in

S. Mittal
Franklin Templeton, Hyderabad, India
e-mail: sukrit.mittal@franklintempleton.com

K. Deb
Michigan State University, East Lansing, MI, USA
e-mail: kdeb@msu.edu
URL: http://www.coin-lab.org

S. Mukhopadhyay et al. (eds.), *Contemporary Advancement in Evolutionary Multi-objective Optimization*, Genetic and Evolutionary Computation,
https://doi.org/10.1007/978-981-95-7908-2_2

and diversity. The authors envision that the synergy between EMâO and ML could be utilized further to enrich each of these domains, for tacking complex real-world problems.

1 Introduction

A multi-objective optimization problem (MOP) can be stated as follows:

$$\begin{aligned} &\text{Minimize} \;\; F(X) = \{f_1(X), f_2(X), \ldots f_M(X)\}^T, \\ &\text{subject to} \;\; X \in \mathcal{X}, \end{aligned} \tag{1}$$

where $\mathcal{X}$ represents the feasible decision (variable) space satisfying all the inequality and equality constraints as applicable; $F : \mathcal{X} \rightarrow R^M$ consists of M real-valued objective functions; and R^M represents the objective-space ($\mathcal{Z}$-space). The *attainable objective set* is defined as $\{F(X)|X \in \mathcal{X}\}$. For any solution $X_j \in \mathcal{X}$, if $F(X_j)$ belongs to the boundary of $F(\mathcal{X})$, then X_j is a Pareto-optimal solution, characterized by the fact that its improvement in any one objective will necessitate sacrificing in some other objective(s). The set of all possible Pareto-optimal solutions constitute the Pareto-set (PS) in $\mathcal{X}$-space and their images in $\mathcal{Z}$-space constitute the Pareto-front (PF). Since a Pareto-optimal solution can not be worse than any other solution in all the objectives simultaneously, it is *not dominated* by any other solution. Hence, the solutions constituting the PF represent the set of non-dominated solutions. More formally, optimal solutions can be defined through the concepts of *partial ordering* [58] and *Pareto-dominance* [43].

The goal for an Evolutionary Multi-objective Optimization Algorithm (or EMOA) is evolve a set of random initial solutions using survival of the fittest based *selection* to approximate the PF, in a single simulation run. The quality of approximation can be judged in terms of: (i) *convergence*—proximity of the EMOA solutions to the true PF, and (ii) *diversity*—extent and uniformity with which the true PF gets covered by the EMOA solutions. While a wider coverage of EMOAs can be found in [9], a brief perspective is presented below. The Vector Evaluated Genetic Algorithm (VEGA) [57] is acknowledged as the first EMOA. Subsequently: (a) proposition of *Pareto-dominance-based ranking* [23] paved way for the first generation EMOAs including MOGA [20], NSGA [61], and NPGA [26], and (b) the notion of *elitism* [10] enabled second generation EMOAs including SPEA [71] and NSGA-II [11]. These EMOAs relied on using Pareto-ranking to pursue *convergence*, and a density-based criterion to pursue *diversity*. However, in mid-2000s, two new EMOA approaches emerged. One of these integrated both *convergence* and *diversity* considerations into a single *indicator* to guide selection [70]. The other approach proposed *decomposition* [66] of an MOP into several single-objective problems, whose solutions approximate PF of the MOP. Several versions of indicator-based and decomposition-based EMOAs have been proposed since then. Notably, the development of EMOAs was also accompanied by indicators for their performance assessment, including Generational Distance (GD) [62] for convergence; Indicator spread (I_s) [11] for diversity; and joint convergence-diversity indicators including Hypervolume (HV) [71] and Inverted Generational Distance (IGD) [62]. For specific test problems, some problem functions are also used to assess performance, e.g., in DTLZ problems [15], the function $g(X)$ having a value close to zero implies close convergence to the PF.

It is critical to note that during 2003–2005, it was alarmingly noted that in problems with $M \geq 4$, the *search* ability of popular EMOAs (pursuing Pareto-ranking based convergence and density-based diversity) severely deteriorates leading to poor PF-approximation [27, 31]. This was attributed to the fact that in such problems, a significant proportion of solutions becomes non-dominated from the early generations [12, 32], effecting: (i) *dominance resistance*—poor selection pressure for convergence, and (ii) *active diversity promotion*—diversification of poorly converged solutions [49]. This necessitated distinguishing problems with $M \geq 4$, as many-objective problems (MaOPs), prompting the development of algorithms capable of handling both MOPs and MaOPs—addressable as Evolutionary Multi- and Many-objective Optimization Algorithms (EMâOAs). Consequently, three broad research directions emerged, including: (i) the *objective reduction* approach [4, 12, 29, 53], (ii) countering dominance resistance through improved dominance principles, such as *hf*-dominance [56], and (iii) countering active diversity promotion by improving the density criterion [1, 37] or using a *reference-vector (RV)-based framework*. In case of the latter, the $\mathcal{Z}$-space is discretized through more or less uniformly distributed RVs to guide convergence and diversity. This approach is manifested by several EMâOAs, including NSGA-III [14, 30], MOEA/D [66], θ-DEA [65], MOEA/DD [35], and LHFiD [56].

Despite great advances over four decades, there are several problem characteristics categorizable as convergence-hardness, diversity-hardness, feasibility-hardness, bias, multi-modality, and inter-variable linkages [7, 15, 19, 39, 41, 72] in the face of which most existing EMâOAs can struggle. Since in real-world problems, there could be a mix of difficult problem characteristics, not known a priori, the authors here found it intriguing to investigate *if* and *how* EMâOAs' generational populations can be subjected to Machine Learning (ML) techniques to address different aspects of optimization, including problem-modeling (pre-search); search efficiency; and post-optimality decision making.

This chapter seeks to present a glimpse of existing research themed around ML-based enhancements in EMâO domain. Understandably, all aspects under this theme can not be covered with equal emphasis. Hence, an overview is presented in one section, while the other sections present in more details the authors' recent efforts to enhance search efficiency of RV-based EMâOAs with an explicit focus on enhancing *convergence* and *diversity* in isolation of each other, and also conjunctly. The choice of RV-based framework is motivated by the fact that evolving solutions become more tractable along and across the RVs. The remainder of this chapter is structured as follows: Sect. 2 provide an overview of ML-based enhancements in EMâO domain, following which Sects. 3, 4, and 5 detail the pro-convergence IP2 operator, the pro-diversity IP3 operator, and the pro-convergence-diversity UIP operator, respectively. Section 6 outlines future research directions, and the chapter concludes with Sect. 7.

2 ML-Based Enhancements in EMâO: An Overview

This section, seeks to present an overview of ML-based enhancements in EMâO. For clarity, it is being structured into subsections focusing on different aspects of optimization, namely, fidelity of problem-model (pre-search), search efficiency, post-optimality analysis, and decision making, respectively.

2.1 ML-Based Enhancements to Gauge Fidelity of Problem-Model

It has been highlighted above that the applicability of popular Pareto-ranking based EMOAs on problems with $M \geq 4$ was becoming increasingly questionable during 2003-2005. In this background, an intriguing proposition was made in 2006 [12], that given a MaOP with an objective set $\mathcal{F}_0 = \{f_1, f_2, \ldots, f_M\}$:

- some objectives could be correlated with others, enabling identification of the redundant objective set ($\mathcal{F}_{\text{redn}}$).
- elimination of $\mathcal{F}_{\text{redn}}$ would point to an *essential* objective set—the smallest set of conflicting objectives ($\mathcal{F}_{\mathcal{T}}, |\mathcal{F}_{\mathcal{T}}| = m$) which can generate the same PF as $\mathcal{F}_0$.

In essence, this approach asserted the importance and benefits of investigating the fidelity of the problem model, by determining whether $m < M$. In that:

- if $m \leq 3 < M$: it implies that a difficult-to-solve MaOP gets reduced to an MOP, that could easily be solved using any second-generation EMOA.
- if $4 \leq m \leq M$: even though $\mathcal{F}_{\mathcal{T}}$ still poses a MaOP, the reduced problem representation offers significant benefits such as higher search efficiency, lower computational cost, and ease in visualization and decision-making.

To realize the above, a Principal Component Analysis (PCA) based approach was integrated with NSGA-II, and tested for *on-the-fly objective reduction* on problems with up to 50 objectives. Its limitations were overcome in [53] by utilizing a non-linear dimensionality reduction technique, namely, Maximum Variance Unfolding (MVU). These approaches culminated in a common ML-based objective reduction framework [52]. Parallelly, other objective reduction approaches based on different principles were also proposed [4, 29, 59].

2.2 ML-Based Enhancements to Augment Search Efficiency

This subsection highlights some of the key studies towards improving the search efficiency, including the innovized repair operator, surrogate modeling, model-based offspring sampling and efficient mutation approaches.

The Innovized Repair (IR) operator is an ML-based technique used within EMâOAs to enhance their search capabilities [21, 22]. This operator leverages the insightful design rules or relationships embedded within trade-off solutions. The process involves extracting these principles (within different variables), often structured as mathematical relationships, from non-dominated solutions discovered in intermediate generations of an EMâOA run. Typically, the learning and repair mechanism is initiated after a certain percentage (e.g., 33%) of the total maximum number of generations ($t_{\max}$) has passed, allowing for some initial stability of the population. In subsequent generations, the learned relationships are applied to repair offspring solutions that are initially created using the bio-inspired operators (crossover and mutation). This repair influences subsequent iterations by guiding the creation of offspring based on the characteristics of good solutions learned from the population.

The surrogate modeling approach in EMâOAs involves using ML methods to learn the variable-objective relationships, locally approximating the potentially computationally expensive function and constraint evaluations. The main purpose is to reduce the number of actual solution evaluations needed for convergence, which is particularly beneficial for real-world problems where obtaining objective and constraint values is costly [2, 8, 17]. Various ML techniques like Gaussian random field models [18], Kriging [60], random forests [63], and support vector regression [28] can be used to model individual objective and constraint functions or groups of functions. In addition to approximating evaluations, surrogate models have been applied to other tasks within an EMâOA, such as facilitating local search [33, 69] and efficient offspring creation [34, 42].

The model-based offspring sampling approach was originally proposed in Estimation of Distribution Algorithms (EDAs) [48], which extracts statistical information from the search space based on elite solutions using ML methods like Bayesian networks or decision trees. A probabilistic model of these elite solutions is built, and then new offspring solutions are generated by sampling from this model. A key advantage is its potential to exploit inter-variable dependencies when generating offspring, particularly useful when such dependencies exist in the PS. However, challenges exist in selecting the elite solutions and choosing the strategy for building the probabilistic model [5]. Some EMâOAs like M-MOEA [68], Hybrid [67], IM-MOEA [6], and GMOEA [24] combine bio-inspired operators with model-based sampling to create offspring solutions.

Mutation is a bio-inspired operator in EMâOAs, primarily employed to maintain diversity within the population [25]. While not always effective on its own, mutation plays a crucial role alongside crossover operators for efficient search [23]. Traditionally, mutation involves modifying one or more variables in a solution with a certain probability, controlling the extent of modification with an index. More efficient approaches have been explored, including the integration of reinforcement learning (RL) to adapt mutation parameters. For instance, NSGA-RL [3] learns suitable mutation parameters for each variable on-the-fly, while RL-NSGA-II [51] uses RL to decide which variable(s) to modify and the extent of modification. Another efficient method involves learning and adapting the search space itself using Principal Component Analysis (PCA); parent solutions are mapped to a reduced space, mutation occurs there, and offspring are mapped back to the original space before evaluation [64].

2.3 ML-Based Enhancements Towards Post-optimality Analysis

In EMâO, generally, a uniform distribution of solutions across PF is desired. However, this cannot always be guaranteed due to the stochasticity involved in EMâOAs and/or difficult problem characteristics. In a contrasting scenario, a higher concentration of solutions in specific parts of PF may be desired by the Decision makers (DMs). To meet such requirements, an ML-based approach has been proposed [13, 55], which treats a given PF-approximation as input and trains an ML model to capture the relationship between pseudo-weight vectors derived from the objective vectors in the PF-approximation (F in $\mathcal{Z}$), and their underlying variable vectors (X in $\mathcal{X}$). This trained ML model is applied to create optimal solutions (X vectors) for different pseudo-weight vectors—chosen to help fill a gap in the available PF-approximation, or to enhance the concentration of solutions in some part of the PF as desired by the DMs. A sample representation for the latter is shown in Fig. 1, on the real-world crashworthiness problem [38], where the concentration of solutions in a particular region of the PF approximation has been enhanced using this approach.

2.4 ML-Based Enhancements to Assist in Decision Making

Decision-Making involves selecting the most desirable solution(s) from the obtained PF-approximation. In the context of MOPs, different methods [43, 50] in practice include: (i)*knee*-point selection, where a knee refers to a point on the PF at which

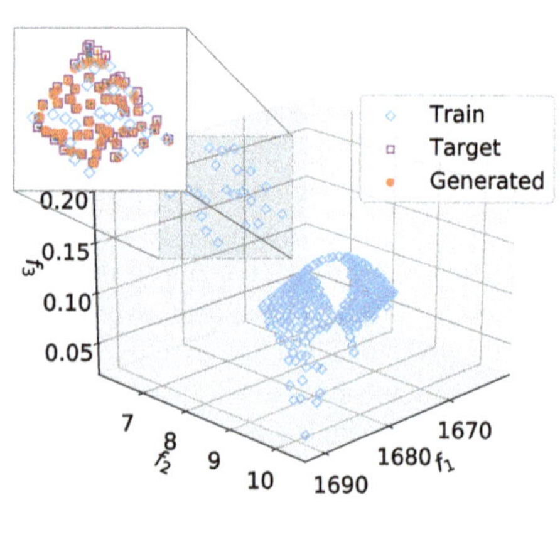

Fig. 1 Additional solutions (shown in red color) are created as part of the post-optimality analysis in regions of low density of solutions on the obtained PF approximation (top left region, shown in blue color) (taken from [55])

the slope changes abruptly, indicating that it offers a fine balance between objective and any transition away from it will imply a minor gain in an objective with large sacrifice in another, (ii) weighting objective functions, where assigning weights to different objectives allows for a more formal comparison of solutions, (iii) Multi-Criteria Decision-Making (MCDM) techniques, such as TOPSIS (Technique for Order Preference by Similarity to an Ideal Solution) where solutions can be rank and selected based on their proximity to an ideal solution, and (iv) Interactive Decision-Making, where the optimization process elicits DMs' preferences interactively to guide the search towards the solutions most preferred by the DMs, as against the whole PF.

In the context of MaOPs, the applicability or effectiveness of the above methods is limited by the fact that—determining knee points is a non-trivial task, and DMs' cognitive constraints while handling many objectives may lead to objective/solution preferences devoid of *objectivity*, *repeatability*, *consistency*, and *coherence*. Towards a rational decision support, the ML-based objective reduction framework [52] (Sect. 2.1) has been extended in [16]. This decision support besides revelation of the essential-objective set, also enables *preference-ranking of objectives* which in turn enables: (i) δ-MOSS analysis: determination of the *smallest* objective set for a user-defined permissible error δ, and (ii) k-EMOSS analysis: determination of k-sized subset associated with minimum error, where k is user-defined. As a representative example, the results from a real-world storm-drainage problem are presented in Fig. 2 (taken from [55]). Each plot compiles 30 analysis corresponding to 30 PF approximations from an EMâOA. The height of the bars therein mark the percentage of runs in which any objective falls in the δ- or k-set for varying δ or k, respectively. This data-driven analysis, aided by visual representation, is designed to help DMs articulate their preferences in a more objective, repeatable, consistent, and coherent manner.

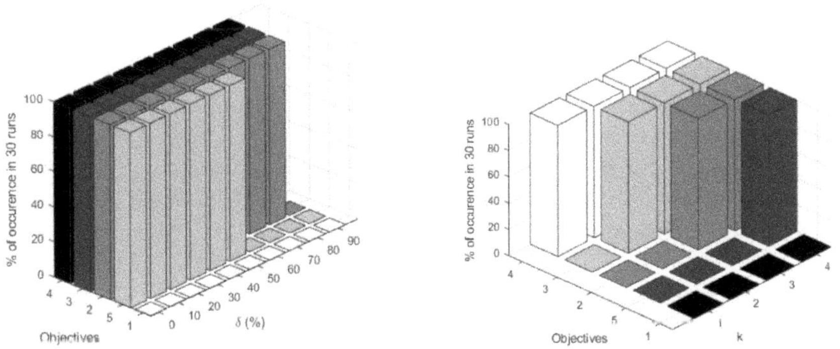

(a) δ-MOSS analyses for varying δs (b) k-EMOSS analyses for varying ks

Fig. 2 Storm Drainage Problem: Graphical illustration of the many-objective decision support proposed in [16]. Each plot compiles 30 analysis corresponding to 30 PF approximations from an EMâOA (taken from [55])

3 Machine Learning to Aid in Faster Convergence

EMâOAs strive to find a well-converged and well-diversified set of solutions approximating the true PF. To enhance the search efficiency of EMâOAs, particularly in accelerating their convergence towards the PF, ML methods have been increasingly explored [40]. This section will focus on the Second Innovized Progress (IP2) operator[1], a notable ML-assisted approach designed to improve the convergence of RV-EMâOAs [46].

The development of the IP2 operator is motivated by the understanding that an RV-EMâOA can be viewed as a time-varying process where the population of solutions gradually migrates towards optimal regions, guided by the RVs. By analyzing the chronological sequence of good (say non-dominated) solutions across generations using ML methods, it is possible to capture efficient search trajectories along the RVs. Traditional RV-EMâOAs relying solely on bio-inspired variation operators, which may exhibit inefficient search behavior when dealing with complex real-world problems. The IP2 operator builds upon the concept of *online innovization* [21, 22], where design rules or efficient search trajectories are learned during the algorithm's execution and are used to guide the search. Specifically, IP2 operator aims to address the convergence-related challenges in RV-EMâOAs by learning from the progress made in previous generations along the pre-defined RVs.

3.1 IP2 Operator

The IP2 operator whose scope is summarized above encapsulates three modules, including: *Training-dataset construction*, *ML training*, and *offspring progression*. In any generation t of an RV-EMâOA, if the IP2 operator is invoked, all these modules are executed sequentially before the offspring solutions are evaluated. The design of these modules is briefly discussed below.

1. Training-Dataset Construction: In any generation t of an RV-EMâOA, this module constructs a training dataset by mapping solutions from an input archive (A_t) to a target archive (T_t) along the respective RVs. T_t stores the best selected solutions found so far along each RV up to the current generation t, whereas A_t contains solutions from a few past generations. The mapping essentially captures the 'progress' made from previous generation solutions in A_t to better solutions in T_t along the RVs in the $\mathcal{Z}$-space, and the corresponding variable vectors in $\mathcal{X}$-space form the training data.
2. ML Training: This module trains a machine learning model on the generated training dataset. The goal is to learn the directional improvements in the $\mathcal{X}$-space that correspond to the observed progress in the $\mathcal{Z}$-space along the RVs. In IP2, RF

[1] The first Innovized Progress (IP) operator used a feedforward artificial neural network with limited success [44, 45].

is used as the ML method due to its computational efficiency compared to Artificial Neural Networks (ANNs) used in earlier innovized progress operator [46]. The trained RF model learns a mapping from an existing solution to a potentially better solution in the $\mathcal{X}$-space.

3. Offspring Progression: This module utilizes the trained RF model to create *pro-convergence* offspring in the current generation. A fraction (e.g., 50%) of the naturally produced offspring (created using crossover and mutation) are randomly selected. The trained RF model is then used to predict the variable vectors of potentially better solutions. By applying this step, these offspring are 'progressed' towards potentially better regions of the $\mathcal{X}$-space, aiming for faster convergence to the PF in $\mathcal{Z}$-space. Keeping this fraction $\leq 50\%$ ensures that at least 50% of the total offspring are always naturally produced, avoiding loss of diversity in solutions in $\mathcal{X}$-space due to over-exploitation by ML models. IP2 also incorporates boundary repair using an inverse parabolic spread distribution to handle variables going out of bounds and a mechanism called 'jutting' to control the extent to which the ML model's predicted direction is followed.

The IP2 operator can be integrated into various RV-EMâOAs, such as NSGA-III, by invoking these modules at certain generations. The detailed implementation of these modules and their methodical integration with an RV-EMâOA can be found in [46, 55]. Notably, the IP2 operator is not invoked in every generation. Its *first* invocation is contingent upon all solutions in the population being non-dominated. The *subsequent* invocations are governed by a frequency parameter called $t_{\text{freq}}^{\text{IP2}}$, which dictates the number of generations between its consecutive invocations. This parameter is adapted on-the-fly based on the survival rate of the offspring progressed by the IP2 operator (Q^{IP2}), aiming to ensure a balance between convergence and diversity, while avoiding ad-hoc parameter fixations. Notably, the first and subsequent invocation criteria, described above, also help mitigate potentially erroneous learning due to noisy and/or misleading search trajectories, especially in the early generations of an RV-EMâOA run.

3.2 *Summary of Experimental Validations*

The performance of IP2 operator, specifically integrated with NSGA-III (referred to as NSGA-III-IP2), has been compared against the base NSGA-III on a range of convergence-hard problems [7, 15, 72]. These experiments compared performance over 31 randomly seeded runs for each test problem, using Wilcoxon ranksum test with p-value threshold of 0.05. The termination generation ($t_{\max}$) was determined on-the-fly for NSGA-III-IP2 using population stabilization [54], and the same $t_{\max}$ was then used for NSGA-III to ensure a fair comparison. The primary performance indicator used was HV, supported by the $g(X)$ function for additional insights into convergence levels. See [55] for the complete analysis.

General Trends: In terms of HV, NSGA-III-IP2 performed statistically better in about 29% instances and better or equivalent in about 95% instances, compared to NSGA-III. However, in terms of $g(X)$, NSGA-III-IP2 performed statistically better in about 56% instances, and better or equivalent in all instances. These $g(X)$ trends indicate a general improvement in convergence levels while the HV trends indicate that in majority of the test problems, diversity was not adversely impacted due to the IP2 operator.

Insights: For problems where the base NSGA-III can approximate the PF well when run sufficiently long, the IP2 operator primarily acts to speed up the PF approximation. For example, on the two-objective Z̃DT1 problem (Fig. 3a), NSGA-III-IP2 showed superior hypervolume in the initial generations, while performed only similar in later generations. However, for harder problems where the base NSGA-III might struggle to approximate the PF well, the IP2 operator can help improve the quality of the PF approximation. For example, on the three-objective MaF12 problem (Fig. 3b), NSGA-III-IP2 showed better hypervolume over the entire run, compared to NSGA-III.

3.3 *Limitations of the IP2 Operator*

Despite the demonstrated benefits of the IP2 operator, it entails some inherent limitations, such as:

- Performance Ceiling: If the base RV-EMâOA already provides a good approximation of the PF on a given problem, then the IP2 operator might only offer improvements in terms of reaching that performance faster, rather than achieving a significantly better final PF approximation.
- Dependence on Population Diversity: The IP2 operator relies on learning meaningful search directions from an inter-generational set of solutions. If the algorithm loses population diversity during the search, the ML model's extrapolatory

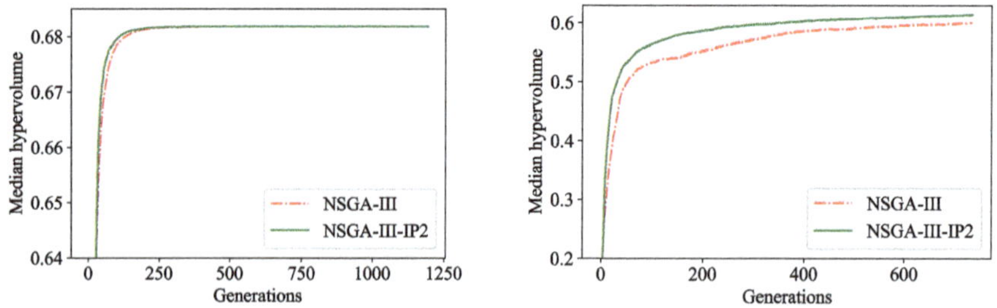

(a) Z̃DT1 (two-objective) (b) MaF12 (three-objective)

Fig. 3 Generation-wise median hypervolume till respective $t_{\max}$ generations (taken from [55])

capabilities might be limited, and IP2 may fail to induce further improvements. Importantly, the IP2 operator in isolation does not have an explicit mechanism to counter a lack of diversity.
- Computational Overhead: While IP2 does not require additional solution evaluations, it does introduce a computational overhead due to the training and application of the ML model. If the time required for a single solution evaluation in the base algorithm is very low compared to the time needed for ML training, the overall computational time might not be reduced, especially on simpler test problems. However, for real-world problems with computationally expensive solution evaluations, the savings in number of generations to reach a certain performance level can outweigh the overhead.

These limitations highlight the importance of considering the diversity aspect alongside convergence enhancement, which has motivated the development of other ML-assisted operators for diversity, and simultaneous convergence and diversity enhancement.

4 Machine Learning to Aid Better Diversity

While many RV-EMâOAs excel at convergence, achieving and maintaining good diversity, in terms of both spread and uniformity, can still be challenging, especially on diversity-hard problems. To address this, ML methods have been explored to enhance the diversity capabilities of RV-EMâOAs. This section will focus on the Third Innovized Progress (IP3) operator, a recently proposed ML-assisted approach designed for diversity enhancement in RV-EMâOAs [47].

The primary motivation behind the IP3 operator stems from the observation that while operators like IP2 effectively improve convergence by learning directional improvements towards the PF along RVs, they are not designed to explicitly enhance the diversity of the obtained solutions. Diversity in the context of PF approximation includes both the extent of spread (coverage of the PF) and the uniformity of the distribution of solutions within that spread. RV-EMâOAs, despite their ability to guide solutions towards different parts of the $\mathcal{Z}$-space using RVs, can still struggle with maintaining a well-spread and uniform set of solutions, particularly in complex $\mathcal{Z}$-spaces [47]. Therefore, the IP3 operator was proposed to specifically learn and induce search trajectories that lead to a better spread and more uniform distribution of solutions across the PF. It operates by learning efficient search trajectories in the $\mathcal{X}$-space based on the mapping of intra-generational solutions in the $\mathcal{Z}$-space, across RVs.

4.1 IP3 Operator

Similar to the IP2 operator, the IP3 operator also encapsulates three modules, including: *Training-dataset construction*, *ML training*, and *offspring creation*. In any generation t of an RV-EMâOA, if the IP3 operator is invoked, all these modules are executed sequentially before the offspring solutions are evaluated. These design of these modules is briefly discussed below.

1. Training-Dataset Construction Module: This module constructs M training datasets (where M is the number of objectives). The goal is to facilitate improvement in each of the M objectives individually to enhance diversity. For a given solution in the parent population (P_t), its mappings to the neighboring solutions associated with different RVs are identified based on their projections onto the $(M - 1)$ unit simplex defined by the RVs. These mappings, from a solution to potentially more diverse solutions, form the training data. The training dataset is designed to cater to both aspects of diversity: spread and uniformity.
2. ML Training Module: This module trains M ML models (one for each of the M training datasets) to learn the directional improvements in the $\mathcal{X}$-space that correspond to the desired diversity enhancement in the $\mathcal{Z}$-space. While the original IP3 operator employed the k-Nearest Neighbors (kNN) as the ML method, it is important to note that other multi-output regression methods could also be used. The M trained ML models are each intended to facilitate improvement in a distinct objective, thereby promoting diversity. The variable bounds are also considered during the training process to guide the learning.
3. Offspring Creation Module: This module utilizes the M trained ML models to create *pro-diversity* offspring in the current generation. A fraction (e.g., 50%) of the total offspring population is generated using the IP3 operator. Similar to IP2 operator, keeping this fraction $\leq 50\%$ ensures that at least 50% of the total offspring are always naturally produced, avoiding loss of diversity in solutions in $\mathcal{X}$-space due to over-exploitation by ML models. This is further divided into two equal parts: one part (e.g., 25% of the total population) aims to improve the spread of solutions by progressing solutions towards the boundaries of the $\mathcal{Z}$-space, and the other part (e.g., 25% of the total population) aims to improve the uniformity of solutions by filling gaps in the current PF approximation. The trained ML models are used to predict beneficial directions of change in the variable vectors of selected parent solutions to achieve these diversity-enhancing objectives. The phenomena is referred to as 'boundary progression' and 'gap progression' for improving spread and distribution respectively, which collectively create these pro-diversity offspring.

The IP3 operator can also be integrated into various RV-EMâOAs, such as NSGA-III, by invoking these modules at certain generations. The detailed implementation of these modules and their methodical integration with an RV-EMâOA can be found in [47, 55]. Similar to the IP2 operator, the IP3 operator is also not invoked in every generation. Unlike IP2, its *first* invocation is contingent upon the population being mildly stabilized [54]. However, the *subsequent* invocations are governed by a frequency parameter called $t_{\text{freq}}^{\text{IP3}}$, which is adapted on-the-fly based on the survival rate of the offspring created by the IP3 operator (Q^{IP3}), similar to the IP2 operator. Similarly, the first and subsequent invocation criteria, described above, also help mitigate potentially erroneous learning due to noisy and/or misleading search trajectories, especially in the early generations of an RV-EMâOA run.

4.2 Summary of Experimental Validations

The performance of IP3 operator, integrated with NSGA-III (referred to as NSGA-III-IP3), has been compared against NSGA-III on a range of diversity-hard problems [19, 39, 41]. The experimental settings and performance indicators used were same as used for the IP2 operator. See [55] for the complete analysis.

General Trends: In terms of HV, NSGA-III-IP3 performed statistically better in about 43% instances and better or equivalent in about 96% instances, compared to NSGA-III. However, in terms of $g(X)$, NSGA-III-IP2 performed statistically better or equivalent in only 75% instances. While the HV trends indicate a general performance improvement, the $g(X)$ trends indicate that the convergence was adversely impacted in the remaining 25% instances. However, the overall HV trend indicates that the diversity improvement through IP3 compensated for the impact on convergence in majority of the instances.

Insights: As can be observed in Fig. 4a, NSGA-III-IP3 clearly achieved a better PF approximation than NSGA-III. Although NSGA-III-IP3 could not achieve the desired spread across the true PF, the improvement in the spread of solutions over NSGA-III could only be attributed to the search efficacy infused by the IP3 operator toward diversity enhancement. Similarly, Fig. 4b clearly shows the better PF

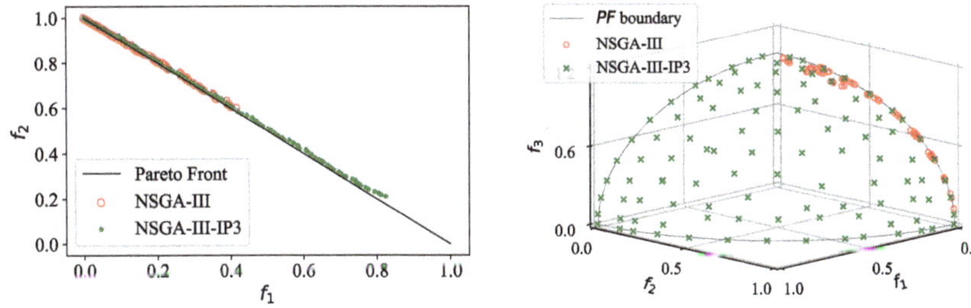

(a) CIBN1 (two-objective) (b) DASCMOP9 (three-objective)

Fig. 4 Final obtained solutions in the median run at respective t_{max} generations (taken from [55]) approximation achieved by NSGA-III-IP3 on DASMOP9 problem, indicating that IP3 operator is suitable for diversity-enhancement in more than two objectives as well.

4.3 Limitations of the IP3 Operator

Despite the demonstrated benefits of the IP3 operator, it entails some inherent limitations, such as:

- Potential Trade-off with Convergence: While designed to complement the convergence of the base RV-EMâOA, on some problems, the strong emphasis on diversity enhancement in the early generations might lead to a temporary slower convergence rate compared to the base algorithm or a convergence-focused operator like IP2. However, final convergence is generally not compromised.
- Computational Overhead: The IP3 operator introduces computational overhead due to the training and application of M ML models in each generation where it is invoked. While it avoids additional function evaluations, the time taken for ML training and offspring creation can be significant, especially for a large number of objectives. The choice of the ML method (e.g., kNN) can influence this overhead.

These limitations highlight the importance of considering both convergence and diversity enhancement simultaneously, which has led to the development of unified operators like UIP that integrate the strengths of both IP2 and IP3.

5 Machine Learning to Aid in Simultaneous Convergence and Diversity Enhancement

As discussed earlier, RV-EMâOAs aim to achieve both convergence to and diversity across the true PF. Previous sections have focused on IP2 and IP3 operators for enhancing convergence and diversity, respectively. NSGA-III integrated with IP2 (NSGA-III-IP2) shows superior convergence on convergence-hard problems, while NSGA-III integrated with IP3 (NSGA-III-IP3) excels in maintaining diversity on diversity-hard problems. These findings underscore the operators' ability to balance convergence and diversity crucial for PF approximation. However, characterizing a problem a priori as convergence-hard or diversity-hard remains challenging. Hence, this section discusses the Unified Innovized Progress (UIP) operator, combining IP2 and IP3 to improve both convergence and diversity without assuming problem characteristics beforehand [47, 55].

5.1 UIP Operator

The UIP operator is not an independent operator which requires a goal-driven training-dataset construction and ML training, leading to offspring creation. Instead, the UIP operator is about invocation of the IP2 and IP3 operators, simultaneously or in isolation of each other, leading to creation of pro-convergence and/or pro-diversity offspring. For the sake of brevity, and RV-EMâOA integrated with the UIP operator is hereon referred to as RV-EMâOA-UIP.

Notably, in any generation t of an RV-EMâOA-UIP, none, one, or both of the IP2 and IP3 operators may be invoked. The first invocation of these constitutive operators is subjected to specific initiation conditions being met (as described in Sects. 3 and 4). Furthermore, their subsequent invocations are guided by the adaptive frequency parameters ($t^{\text{IP2}}_{\text{freq}}$ and $t^{\text{IP3}}_{\text{freq}}$), which represent the number of generations between two successive invocations for each operator. Four different scenarios are possible, depending on which of the initiation conditions are met and the values of the corresponding frequency parameters. First, the most comprehensive scenario where both IP2 and IP3 are invoked in an RV-EMâOA-UIP generation, is discussed below, with reference to Fig. 5. In that scenario:

- First 100% (N) natural offspring are created, denoted by Q^{V} (in boxes (a) and (b) under 'Offspring Population').
- IP2 creates 50% pro-convergence offspring Q^{IP2} (in box (c)), through the progression of 50% of randomly chosen Q^{V}.
- IP3 creates 50% pro-diversity offspring Q^{IP3} (in box (d)), through the progression of 50% of the judiciously chosen parent solutions.
- The new offspring Q^{IP2} and Q^{IP3} (in boxes (c) and (d), collectively of size N) are combined with the parent population, leading to the combined population, of size $2N$. Notably, the *extra* offspring, in boxes (a) and (b), are not evaluated.

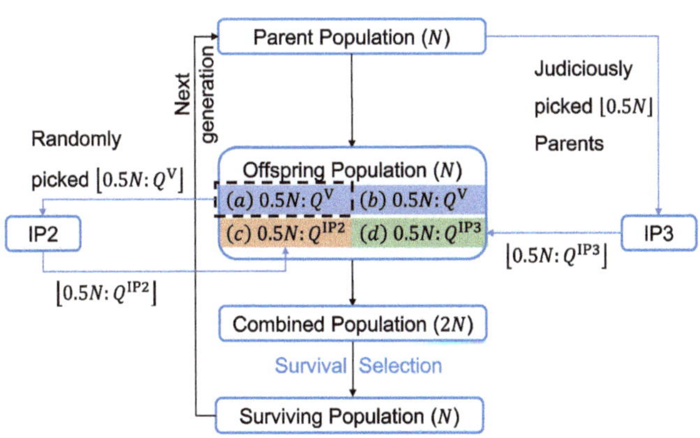

Fig. 5 A schematic for an RV-EMâOA-UIP generation (taken from [47])

- Finally, the survival selection of the underlying algorithm is executed, leading to the surviving population (of size N), which serves as the parent population for the next generation.

Figure 5 also helps to interpret the other scenarios, where in a generation t of an RV-EMâOA-UIP, none or one of the IP2 and IP3 operators may be invoked. For example:

- If in an RV-EMâOA-UIP generation, neither IP2 nor IP3 is invoked: this generation is implemented as the base RV-EMâOA generation. In that, Q^{V}, of size N (boxes (a) and (b)) constitute the final offspring.
- If in an RV-EMâOA-UIP generation, only IP2 gets invoked: Q^{V} of size N (boxes (a) and (b)) are created and half of them randomly chosen (box (a)) are subjected to the IP2 operator. The newly created Q^{IP2} (box (c)) along with the unused Q^{V} (box (b)) constitute the final offspring.
- If in an RV-EMâOA-UIP generation, only IP3 is invoked: only half of Q^{V} (box (a)) are created and along with the newly created Q^{IP3} (box (d)), constitute the final offspring.

The enabling steps in an RV-EMâOA-UIP generation, highlighted above, are guided by the need to ensure that: (i) neither convergence nor diversity gets over-emphasized at the cost of the other, (ii) the performance of the base RV-EMâOA is not significantly impacted in a detrimental sense, if the learning (through underlying ML models) happens to be erroneous, for some reason, (iii) no extra solution evaluations vis-à-vis the base RV-EMâOA are incurred, and (iv) *ad-hoc* fixation of introduced parameters in avoided, as far as possible. More details on how each of these conditions are catered to can be found in [47, 55].

5.2 Summary of Experimental Validations

The performance of UIP operator has first been compared with IP2 and IP3 on convergence- and diversity-hard problems, respectively, using NSGA-III as the base algorithm. Subsequently, the integration of UIP operator with multiple RV-EMâOAs has been compared with their respective base algorithms on a wide range of problems, including convergence- and diversity-hardness, and multi- and many-objective problems. See [55] for complete analysis.

Comparison with IP2 and IP3 Operators: The integration of UIP operator with NSGA-III is referred to as NSGA-III-UIP. In the context of IP2 operator on convergence-hard problems, NSGA-III-UIP performs statistically better in about 19% instances and statistically better or equivalent in about 95% instances, compared to NSGA-III-IP2. This implies a general improvement in overall performance with UIP, compared to IP2. A representative hypervolume plot on $\tilde{\text{Z}}$DT6 problem is shown in Fig. 6a, clearly demonstrating the better performance of the UIP operator.

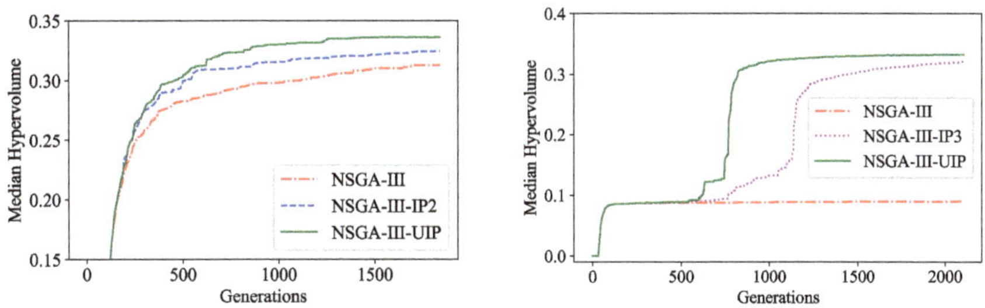

(a) $\tilde{Z}$DT6 (convergence-hard) (b) DASCMOP1 (diversity-hard)

Fig. 6 Generation-wise median hypervolume till respective $t_{\max}$ generations (taken from [55])

Similarly, in the context of IP3 operator on diversity-hard problems, NSGA-III-UIP performs statistically better in about 29% instances and statistically better or equivalent in about 96% instances, compared to NSGA-III-IP3. This implies a general improvement in the overall performance with UIP, compared to IP3. A representative hypervolume plot on DASCMOP1 problem is shown in Fig. 6b, clearly demonstrating the better performance of the UIP operator.

Comparison with Base Algorithm on Both Multi- and Many-Objective Problems: Towards a broader analysis of the performance of the UIP operator, it is additionally integrated with other RV-EMâOAs, namely θ-DEA [65], MOEA/DD [36] and LHFiD [56]. The resulting algorithms are referred to as θ-DEA-UIP, MOEA/DD-UIP and LHFiD-UIP, respectively. The implementation details of these algorithms are described in [55]. To capture the complete picture, NSGA-III results have also been included along with the above algorithms in the following summary.

- On multi-objective problems, including both convergence- and diversity-hard, the UIP variants are statistically better in about 42% instances, and statistically better or equivalent in about 96% instances, compared to their respective base variants, as shown in Table 1.
- On many-objective problems, covering 5, 8 and 10 objectives of the scalable test suites [7, 15], the UIP variants are statistically better in about 28% instances, and statistically better or equivalent in about 89% instances, compared to their respective base variants, as shown in Table 2.

5.3 Limitations of the UIP Operator

The UIP operator offers a significant advancement in multi- and many-objective optimization by effectively integrating the convergence-enhancing capabilities of IP2 and the diversity-enhancing capabilities of IP3. Experimental results on a wide range

Table 1 HV based performance comparison of NSGA-III-UIP, θ-DEA-UIP, MOEA/DD-UIP and LHFiD-UIP with their respective base variants, i.e., NSGA-III, θ-DEA, MOEA/DD and LHFiD, respectively, on convergence-hard (Z̃DT, DTLZ, MaF) and diversity-hard (CIBN, DASCMOP, MW) test suites. Each column shows the median HV (from 31 runs) at the end of t_{max} generations, determined on-the-fly for the respective UIP variants. The symbols "+", "=" or "–" against each variant highlight where these are statistically better than, comparable to, or worse than the UIP variant, respectively. Wilcoxon ranksum test with p-value threshold of 0.05 was used for the statistical comparisons (taken from [47])

Problem	M	t_{max}	NSGA-III	NSGA-III-UIP	t_{TM}	θ-DEA	θ-DEA-UIP	t_{TM}	MOEA/DD	MOEA/DD-UIP	t_{TM}	LHFiD	LHFiD-UIP
Z̃DT1	2	1198	0.681860=	0.681859	1161	0.68128=	0.681299	1155	0.681215=	0.681193	683	0.681874+	0.681871
Z̃DT2	2	1267	0.348794=	0.348794	1237	0.347854=	0.347861	1247	0.347746=	0.347800	774	0.348792 –	0.348793
Z̃DT3	2	1007	1.068445=	1.068492	977	1.067375=	1.067561	1005	1.067135=	1.067364	664	0.987924 –	1.022636
Z̃DT4	2	1767	0.681859=	0.681860	1744	0.681257=	0.681255	1675	0.681196=	0.681192	1481	0.681873+	0.681782
Z̃DT6	2	1836	0.312752 –	0.336457	1822	0.313645 –	0.334052	1771	0.313384 –	0.332646	1209	0.309905 –	0.333011
DTLZ1	3	1497	1.221639 –	1.222575	1118	1.213503+	0.873703	1136	1.216124=	1.202253	633	0.012268=	0.002838
DTLZ2	3	978	0.667327=	0.667323	963	0.651611=	0.651758	912	0.653244=	0.652884	499	0.659976=	0.660592
DTLZ3	3	1750	0.652251 –	0.662523	1237	0.611978=	0.085917	981	0.551322+	0.000000	635	0.000000=	0.000000
DTLZ4	3	1449	0.667309 –	0.667346	1935	0.652918=	0.654825	1226	0.654143 –	0.655937	908	0.661133=	0.661058
MaF1	3	608	0.235973=	0.235082	618	0.228233=	0.229025	625	0.229522=	0.227796	546	0.208156 –	0.209422
MaF2	3	486	0.396887=	0.396411	515	0.390904=	0.391698	513	0.393007=	0.393700	488	0.394240=	0.394093
MaF3	3	2135	1.193651 –	1.194107	2123	1.193223 –	1.194434	2133	1.191562 –	1.194840	1527	1.191826=	1.194385
MaF4	3	1214	0.612050 –	0.629769	1300	0.610459 –	0.620879	1295	0.613593 –	0.617812	736	0.568126=	0.013983
MaF5	3	1345	1.227613=	1.227604	1831	1.228640=	1.228654	1503	1.228484=	1.228521	1119	1.222987 –	1.228257
MaF7	3	1201	0.375791=	0.376159	1166	0.371659=	0.372502	1118	0.371731=	0.371782	1049	0.363208 –	0.371374
MaF8	3	1244	0.463948 –	0.466653	1532	0.000331=	0.000170	1608	0.000109=	0.000117	1552	0.458162=	0.458045
MaF9	3	1160	0.626751=	0.626729	1155	0.610583 –	0.614858	1227	0.610000 –	0.613124	777	1.160416=	1.160459
MaF10	3	996	0.528291=	0.519436	1019	0.477226 –	0.492448	1025	0.508591=	0.499053	944	0.568694=	0.571757
MaF11	3	993	0.980497=	0.979771	1071	1.144277=	1.138350	998	1.142115=	1.142002	800	1.146045 –	1.159070
MaF12	3	725	0.599802 –	0.612526	743	0.534650 –	0.590104	761	0.535199 –	0.592456	626	0.533782 –	0.543327
MaF13	3	1065	0.365532 –	0.377158	1021	0.366427 –	0.369516	1174	0.501206 –	0.524071	946	0.401748 –	0.456261

(continued)

Table 1 (continued)

Problem	M	t_{max}	NSGA-III	NSGA-III-UIP	t_{TM}	θ-DEA	θ-DEA-UIP	t_{TM}	MOEA/DD	MOEA/DD-UIP	t_{TM}	LHFiD	LHFiD-UIP
CIBN1	2	1365	0.327867 –	0.504742	1367	0.335619 –	0.508695	1325	0.334598 –	0.488058	518	0.381507 –	0.506510
CIBN2	2	789	0.658712 –	0.669583	811	0.667394 –	0.670158	766	0.668834 –	0.669695	406	0.670678 –	0.677235
CIBN3	2	960	0.213584 –	0.226977	1013	0.213996 –	0.225995	914	0.214547 –	0.226747	426	0.213834 –	0.222645
CIBN4	3	439	0.912600 –	0.920335	505	0.895468 –	0.898464	480	0.896755=	0.896848	393	0.992818 –	1.028556
CIBN5	3	294	0.629637+	0.628892	299	0.619556=	0.619234	305	0.622079=	0.621281	359	0.622345=	0.622445
DASCMOP1	2	2101	0.089766 –	0.332737	1889	0.087565 –	0.332968	1944	0.132878 –	0.333200	1929	0.082801 –	0.329137
DASCMOP2	2	1944	0.414610 –	0.666735	1851	0.415182 –	0.664793	1963	0.472963 –	0.669519	1876	0.410152 –	0.663136
DASCMOP3	2	1658	0.391774 –	0.399082	1836	0.398531 –	0.489664	1866	0.408387 –	0.483395	2973	0.402701=	0.392024
DASCMOP4	2	1993	0.336866 –	0.336960	2006	0.335362 –	0.335532	2183	0.335677=	0.335616	1741	0.267030=	0.250359
DASCMOP5	2	2113	0.672667=	0.672663	2056	0.671229=	0.671456	2293	0.671290=	0.671308	1863	0.646158=	0.629977
DASCMOP6	2	2432	0.549906 –	0.574797	2536	0.574568 –	0.574866	2024	0.575065 –	0.575132	1887	0.549128=	0.532582
DASCMOP7	3	1731	1.027245=	1.025053	1757	1.014610=	1.011353	1888	1.016219=	1.017115	2257	1.028625=	1.028417
DASCMOP8	3	1675	0.630175=	0.628234	1751	0.648435=	0.632479	1850	0.651319=	0.651529	2115	0.655370=	0.655050
DASCMOP9	3	1537	0.346493 –	0.647739	1524	0.451661 –	0.643508	1511	0.638729 –	0.647314	2201	0.641902 –	0.642864
MW1	2	1042	0.415304=	0.415353	1022	0.414481=	0.414439	1074	0.413709 –	0.414543	1268	0.415326+	0.415005
MW2	2	848	0.483094=	0.491251	833	0.482597 –	0.489976	891	0.482312 –	0.490426	813	0.327672=	0.330055
MW3	2	868	0.469880=	0.469568	1050	0.469747=	0.469411	971	0.469695=	0.469723	1619	0.417336=	0.000000
MW4	3	743	1.041376=	1.041269	756	1.024349=	1.026601	755	1.024167 –	1.028355	879	1.039593=	1.039663
MW5	2	1783	0.083010 –	0.198466	1632	0.100401 –	0.200033	1453	0.195836=	0.200113	1125	0.082853=	0.010235
MW6	2	1244	0.298309 –	0.317611	1212	0.297263 –	0.310665	1272	0.296733 –	0.309847	814	0.119352=	0.119553
MW7	2	906	0.366397=	0.366207	905	0.365294=	0.365080	890	0.365752=	0.366039	814	0.363979=	0.364093
MW8	3	732	0.626663=	0.627444	732	0.609678=	0.612839	668	0.615115=	0.624875	1223	0.391247=	0.472479
MW9	2	1063	0.293673 –	0.295451	986	0.293277 –	0.295267	1128	0.293964=	0.293913	764	0.295782=	0.295258
MW10	2	1071	0.247084=	0.268710	1032	0.246414=	0.263219	1054	0.261012=	0.289881	1680	0.000000=	0.000000
MW11	2	976	0.268232+	0.257800	987	0.265164=	0.208145	1087	0.265863=	0.254258	681	0.268410=	0.268541
MW12	2	1055	0.570528 –	0.570699	988	0.570107 –	0.570262	1085	0.570166 –	0.570440	732	0.570348=	0.570297
MW13	2	1001	0.328966 –	0.346489	990	0.322834 –	0.336527	945	0.322475=	0.334442	723	0.299718=	0.301507
MW14	3	848	0.152752=	0.154984	894	0.156222=	0.155941	960	0.164248=	0.168807	679	0.213535+	0.209878
Total (+/=/ –) ⟶			02/23/**24**			01/25/**23**			01/28/**20**			04/29/**16**	of 49 probs

Table 2 HV based performance comparison of NSGA-III-UIP, θ-DEA-UIP, MOEA/DD-UIP and LHFiD-UIP with their respective base variants, i.e., NSGA-III, θ-DEA, MOEA/DD and LHFiD, respectively. For each test instance of $M = 5$, 8 and 10, median hypervolume is shown at the end of t_{max} generations determined on-the-fly for the respective UIP variants. The symbols "+", "=" or " – " against each base variant highlight where these are statistically better than, comparable to, or worse than the UIP variant, respectively. Wilcoxon ranksum test with p-value threshold of 0.05 was used for the statistical comparisons. (taken from [47])

Problem	M	t_{max}	NSGA-III	NSGA-III-UIP	t_{TM}	θ-DEA	θ-DEA-UIP	t_{TM}	MOEA/DD	MOEA/DD-UIP	t_{TM}	LHFiD	LHFiD-UIP
DTLZ1	5	1367	2.486820=	2.487010	748	2.157168=	2.155684	883	2.670244=	2.456018	859	4.207928=	4.052889
	8	1604	9.988500=	9.988620	875	9.981730=	9.979700	853	6.960960+	5.976893	869	9.988074=	9.982169
	10	2007	17.757700=	17.757700	833	0.000000=	0.000000	808	16.139124+	15.307207	1223	17.757704=	17.757702
DTLZ2	5	860	2.172040=	2.172090	941	2.121200=	2.117510	894	2.121840=	2.120590	578	3.891166=	3.890823
	8	777	9.816830 –	9.818580	905	9.763760+	9.757400	839	9.758620=	9.755060	644	9.847959=	9.847937
	10	793	17.665800 –	17.667500	877	17.591500=	17.592100	834	17.574200=	17.566900	676	17.694772=	17.694515
DTLZ3	5	1071	1.938460+	0.000000	765	2.082338=	2.012005	904	0.260120=	0.066750	776	0.783335+	0.000000
	8	1537	9.741440=	9.759030	847	8.546232+	7.622917	803	8.319002=	8.116630	863	9.245672=	0.014920
	10	1718	17.650300 –	17.664100	819	0.000000=	0.000000	814	9.385090+	6.920623	881	17.633675=	17.105152
DTLZ4	5	912	2.173240=	2.173190	969	2.129440=	2.128680	960	2.133190=	2.130790	868	3.895547=	3.895753
	8	825	9.826530+	9.826270	857	9.731910+	9.707990	784	9.713380=	9.714600	841	9.849535=	9.849960
	10	820	17.677500=	17.677200	763	17.219400=	17.165100	754	17.296200=	17.248200	857	17.696040=	17.696277
MaF1	5	843	0.059222=	0.059753	844	0.056371=	0.056379	856	0.055339=	0.056772	828	0.037931=	0.037928
	8	989	0.015552=	0.015751	928	0.013510=	0.012873	839	0.013111=	0.013142	886	0.000128=	0.000129
	10	830	0.002220=	0.002241	773	0.001562 –	0.001647	757	0.001787 –	0.001837	862	0.000724=	0.000724
MaF2	5	359	0.997132=	0.998834	366	0.988566 –	0.995460	376	0.981635=	0.984138	468	0.867556+	0.843094
	8	422	4.352760 –	4.377980	454	4.122040 –	4.177970	455	4.039790 –	4.096930	580	1.290279+	1.175563
	10	487	7.846630=	7.856730	521	7.475370 –	7.613500	535	7.321450 –	7.494960	594	8.161162+	7.354750

(continued)

Table 2 (continued)

Problem	M	t_{max}	NSGA-III	NSGA-III-UIP	t_{TM}	θ-DEA	θ-DEA-UIP	t_{TM}	MOEA/DD	MOEA/DD-UIP	t_{TM}	LHFiD	LHFiD-UIP
MaF3	5	2238	2.488190 –	2.488200	1209	1.977794=	1.866041	1842	2.483580=	2.378010	1729	2.160181=	2.160109
	8	3333	9.988720 –	9.988720	612	0.000000=	0.000007	1504	0.000000=	0.000045	1453	3.432204+	3.432000
	10	3765	17.757700 –	17.757700	461	0.000000=	0.000000	568	0.000000=	0.000237	1401	17.757727+	17.757130
MaF4	5	604	0.117321=	0.083836	863	0.192197=	0.220878	820	0.154150=	0.168207	1101	0.112496=	0.108572
	8	874	0.118441=	0.153381	963	0.032738=	0.045256	833	8.957987=	8.748836	1172	0.000273 –	0.000533
	10	1069	0.064231 –	0.077138	1222	0.026055 –	0.034063	975	0.015330=	0.019916	1146	0.000000=	0.000000
MaF5	5	1109	2.487950=	2.487940	1157	2.487720=	2.487770	1173	2.487690 –	2.487740	1039	2.160346 –	2.161071
	8	1160	9.988720 –	9.988720	1254	9.988720=	9.988720	1167	9.988720 –	9.988720	970	3.431163 –	3.432212
	10	1196	17.757700 –	17.757700	1292	17.757700=	17.757700	1090	17.757700=	17.757700	1020	15.021723+	6.174841
MaF7	5	1087	0.970398=	0.969063	1090	0.956168=	0.955427	1126	0.968569+	0.964265	1244	0.657965 –	0.716870
	8	912	4.402240=	4.390560	1022	4.222140 –	4.262750	959	4.291200=	4.295080	1562	0.453748 –	0.562075
	10	849	7.602060=	7.595800	766	6.468070=	6.712380	514	4.935440 –	5.351250	1715	0.000000=	0.000000
MaF8	5	1546	0.475295=	0.476568	1767	0.000233 –	0.000340	1853	0.000094 –	0.000208	1675	0.000392+	0.000390
	8	1315	0.709058+	0.686091	1429	0.000517 –	0.000570	1508	0.000390=	0.000382	1647	0.000003=	0.000003
	10	1214	0.577736=	0.572205	1317	0.000075=	0.000081	1438	0.000063=	0.000061	1595	0.000117=	0.000117
MaF9	5	985	0.027410=	0.027633	2037	0.028336=	0.028005	1607	0.026104=	0.027067	5632	0.022343 –	0.024079
	8	716	0.000518 –	0.002425	849	0.000833=	0.001078	745	0.000000=	0.000004	1132	0.000054=	0.000054
	10	774	0.000106=	0.000136	786	0.000014=	0.000042	912	0.000022 –	0.000071	1151	0.000504=	0.000504
MaF10	5	1077	0.939567 –	0.947997	1049	0.830748 –	0.877855	975	0.828694 –	0.879025	1180	0.967104 –	0.991590
	8	853	3.240900 –	3.456380	888	2.709280 –	3.293280	931	2.599670 –	3.326120	1284	1.498149 –	1.578252
	10	855	6.178800+	5.927050	962	5.286690 –	5.609290	938	4.797120 –	5.624200	1475	0.689354+	0.254470

(continued)

Table 2 (continued)

Problem	M	t_{max}	NSGA-III	NSGA-III-UIP	t_{TM}	θ-DEA	θ-DEA-UIP	t_{TM}	MOEA/DD	MOEA/DD-UIP	t_{TM}	LHFiD	LHFiD-UIP
MaF11	5	968	2.437800=	2.437950	1048	2.376760=	2.370580	1020	2.361810=	2.364970	963	2.143115 –	2.147966
	8	1369	9.807760 –	9.847950	1105	9.448400=	9.465670	1073	9.368320=	9.367990	1027	3.395103=	3.396399
	10	1606	17.526500 –	17.613100	1193	16.919800 –	17.053500	1149	16.695000 –	16.906100	1055	15.582983+	14.025161
MaF12	5	573	1.775930=	1.776530	558	1.671030=	1.647670	505	1.637250=	1.622540	663	1.615504 –	1.668499
	8	448	7.256190=	7.254460	431	6.450900=	6.438770	462	5.856190=	5.954880	718	2.724894 –	2.840152
	10	483	13.127900=	13.199400	490	11.461000=	11.612800	496	10.449200=	10.762600	750	0.000000=	0.000000
MaF13	5	820	0.225160+	0.206183	857	0.248616=	0.513728	944	0.409027 –	0.515826	1815	0.204676 –	0.521277
	8	872	0.234528=	0.224742	966	0.269744=	0.297863	1132	0.621752=	1.007820	2091	0.057134 –	0.057623
	10	867	0.172316=	0.165359	1211	0.160839 –	0.213371	1164	1.268000 –	1.918400	2414	0.256875 –	0.474652
Total (+/=/ –) ⟶			05/28/**15**			03/32/**13**			04/30/**14**			10/24/**14**	of 48 probs.

of test problems demonstrate its potential to improve the performance of various RV-EMâOAs simultaneously in terms of convergence and diversity, without incurring additional computational cost in terms of function evaluations. While limitations such as computational cost of IP2 and IP3, and parameter sensitivity of the underlying ML methods exist, the UIP operator represents a promising direction for developing more robust and efficient RV-EMâOAs for challenging real-world optimization problems.

6 Future Extensions

The potential future extensions for ML-assisted EMâO cover several key areas. One major direction involves extending ML interventions beyond the Innovized Progress operators (IP2, IP3, and UIP) to other fundamental components of EMâOAs, including, but not limited to, RV *creation* and *update*; *initialization* of solutions; *normalization* of the $\mathcal{Z}$-space; and algorithm *termination*. Additionally, ML could be utilized for hyper-parameter learning to optimize EMâOA settings. While the book primarily discusses the innovized progress operators in the context of RV-EMâOAs, future research could investigate applying these concepts to other types of EMâOAs that do not rely on RVs. Testing the effectiveness of ML-based interventions on a broader spectrum of real-world applications, particularly combinatorial optimization problems known for their complexity, could also be considered as an important direction for future research.

Further advancements could involve incorporating more sophisticated ML methods, such as ML-assisted Transfer Learning, which allows the knowledge gained from solving one problem to be applied to related problems, offering multi-pronged benefits for the EMâO domain. Regardless of the specific application, future ML-based enhancements must adhere to critical considerations to ensure success, by maintaining the essential convergence-diversity balance of the base EMâOA and favorably managing the ML-based risk-reward trade-off. It is also crucial to avoid introducing extra solution evaluations and ensure that these ML-assisted operations are computationally efficient, operating within reasonable run-time. The template laid down in [55], structured around these core considerations, can guide future research efforts in this promising area.

7 Conclusions

This chapter has explored ML-assisted RV-EMâOAs, highlighting the development and validation of the Innovized Progress (IP) operators. The Second Innovized Progress (IP2) operator, using ML to learn inter-generational progress, has shown to aid faster convergence, improving metrics like hypervolume (HV) and $g(X)$ values. To address diversity, the Third Innovized Progress (IP3) operator has employed ML to learn intra-generational search trajectories, demonstrating improved HV and better PF approximation on diversity-hard problems. Recognizing the difficulty in classifying problems beforehand, the Unified Innovized Progress (UIP) operator has combined IP2 and IP3 functionalities, showing statistically better or equivalent performance in a large majority of instances compared to base algorithms and individual IP operators on both convergence- and diversity-hard problems across multi- and many-objective settings. These operators, while introducing some computational overhead and parameter sensitivity, significantly advance the capability of RV-EMâOAs by simultaneously enhancing convergence and diversity without additional solution evaluations, motivating further studies in this intersecting area of ML and EMâOA.

Acknowledgements Authors thank financial and facilitation support by Government of India under the SPARC project (P66) for this study. Authors also thank Prof. Erik D. Goodman for his expert suggestions in executing this study. Further ideas on the use of ML in improving the performance of EMâOAs can be found in a recent book entitled 'Machine Learning Assisted Evolutionary Multi- and Many-Objective Optimization' [55].

References

1. Adra, S.F., Fleming, P.J.: Diversity management in evolutionary many-objective optimization. IEEE Trans. Evol. Comput. **15**(2), 183–195 (2011). https://doi.org/10.1109/TEVC.2010.2058117
2. Bhattacharjee, K.S., Isaacs, A., Ray, T.: Multi-objective optimization using an evolutionary algorithm embedded with multiple spatially distributed surrogates. In: Multi-Objective Optimization. pp. 135–155. World Scientific (2017). https://doi.org/10.1142/9789813148239_0005
3. Bora, T.C., Mariani, V.C., dos Santos Coelho, L.: Multi-objective optimization of the environmental-economic dispatch with reinforcement learning based on non-dominated sorting genetic algorithm. Appl. Therm. Eng. **146**, 688–700 (2019). https://doi.org/10.1016/j.applthermaleng.2018.10.020
4. Brockhoff, D., Zitzler, E.: Are all objectives necessary? On dimensionality reduction in evolutionary multiobjective optimization. In: Runarsson, T.P., Beyer, H.G., Burke, E., Merelo-Guervós, J.J., Whitley, L.D., Yao, X. (eds.) Parallel Problem Solving from Nature—PPSN IX, pp. 533–542. Springer, Berlin Heidelberg, Berlin, Heidelberg (2006)
5. Chen, Y., Zhang, Y., Abraham, A.: Estimation of distribution algorithm for optimization of neural networks for intrusion detection system. In: Rutkowski L., Tadeusiewicz R., Zadeh L.A., Żurada J.M. (eds) Artificial Intelligence and Soft Computing—ICAISC 2006. ICAISC 2006. Lecture Notes in Computer Science. vol. 4029. Springer, Berlin, Heidelberg (2006)
6. Cheng, R., Jin, Y., Narukawa, K., Sendhoff, B.: A multiobjective evolutionary algorithm using Gaussian process-based inverse modeling. IEEE Trans. Evol. Comput. **19**(6), 838–856 (2015). https://doi.org/10.1109/TEVC.2015.2395073
7. Cheng, R., Li, M., Tian, Y., Zhang, X., Yang, S., Jin, Y., Yao, X.: A benchmark test suite for evolutionary many-objective optimization. Complex Intell. Syst. **3**, 67–81 (2017). https://doi.org/10.1007/s40747-017-0039-7
8. Chugh, T., Jin, Y., Miettinen, K., Hakanen, J., Sindhya, K.: A surrogate-assisted reference vector guided evolutionary algorithm for computationally expensive many-objective optimization. IEEE Trans. Evol. Comput. **22**(1), 129–142 (2018)
9. Coello, C.A.C., Brambila, S.G., Gamboa, J.F., Tapia, M.G.C., Gómez, R.H.: Evolutionary multiobjective optimization: open research areas and some challenges lying ahead. Complex Intell. Syst. **6**, 221–236 (2020). https://doi.org/10.1007/s40747-019-0113-4

10. Deb, K.: Multi-objective optimization using evolutionary algorithms. Wiley, Chichester (2001)
11. Deb, K., Pratap, A., Agarwal, S., Meyarivan, T.: A fast and elitist multiobjective genetic algorithm: NSGA-II. IEEE Trans. Evol. Comput. **6**(2), 182–197 (2002). https://doi.org/10.1109/4235.996017
12. Deb, K., Saxena, D.K.: Searching for Pareto-optimal solutions through dimensionality reduction for certain large-dimensional multi-objective optimization problems. In: IEEE Congress on Evolutionary Computation, pp. 3353–3360 (2006)
13. Deb, K., Gondkar, A., Anirudh, S.: Learning to predict Pareto-optimal solutions from pseudo-weights. In: Emmerich, M., Deutz, A., Wang, H., Kononova, A.V., Naujoks, B., Li, K., Miettinen, K., Yevseyeva, I. (eds.) Evolutionary Multi-Criterion Optimization, pp. 191–204. Springer, Cham (2023)
14. Deb, K., Jain, H.: An evolutionary many-objective optimization algorithm using reference-point-based nondominated sorting approach, Part I: Solving problems with box constraints. IEEE Trans. Evol. Comput. **18**(4), 577–601 (2014). https://doi.org/10.1109/TEVC.2013.2281535
15. Deb, K., Thiele, L., Laumanns, M., Zitzler, E.: Scalable test problems for evolutionary multiobjective optimization. In: Evolutionary Multiobjective Optimization: Theoretical Advances and Applications, pp. 105–145. Springer London, London (2005). https://doi.org/10.1007/1-84628-137-7_6
16. Duro, J.A., Saxena, D.K., Deb, K., Zhang, Q.: Machine learning based decision support for many-objective optimization problems. Neurocomputing **146**, 30–47 (2014). https://doi.org/10.1016/j.neucom.2014.06.076
17. Dutta, S., Gandomi, A.H.: Surrogate model-driven evolutionary algorithms: theory and applications. In: Evolution in Action: Past, Present and Future: A Festschrift in Honor of Erik D. Goodman, pp. 435–451. Springer, Cham (2020)
18. Emmerich, M., Giannakoglou, K.C., Naujoks, B.: Single and multiobjective evolutionary optimization assisted by Gaussian random field metamodels. IEEE Trans. Evol. Comput. **10**(4), 421–439 (2006)
19. Fan, Z., Li, W., Cai, X., Li, H., Wei, C., Zhang, Q., Deb, K., Goodman, E.: Difficulty adjustable and scalable constrained multiobjective test problem toolkit. Evol. Comput. **28**(3), 339–378 (2020). https://doi.org/10.1162/evco_a_00259
20. Fonseca, C.M., Fleming, P.J.: Genetic algorithms for multiobjective optimization: formulation, discussion and generalization. In: Forest, S. (ed.) Fifth International Conference on Genetic Algorithms, pp. 416–423 (1993)
21. Gaur, A., Deb, K.: Adaptive use of *innovization* principles for a faster convergence of evolutionary multi-objective optimization algorithms. In: Proceedings of the 2016 on Genetic and Evolutionary Computation Conference Companion. p. 75–76. GECCO '16 Companion, Association for Computing Machinery, New York, NY, USA (2016). https://doi.org/10.1145/2908961.2909019
22. Gaur, A., Deb, K.: Effect of size and order of variables in rules for multi-objective repair-based *innovization* procedure. In: 2017 IEEE Congress on Evolutionary Computation (CEC), pp. 2177–2184 (2017). https://doi.org/10.1109/CEC.2017.7969568
23. Goldberg, D.E.: Genetic algorithms in search, optimization, and machine learning. Addison-Wesley, Boston, MA (1989)
24. He, C., Huang, S., Cheng, R., Tan, K.C., Jin, Y.: Evolutionary multiobjective optimization driven by generative adversarial networks (GANs). IEEE Trans. Cybern. **51**(6), 3129–3142 (2021). https://doi.org/10.1109/TCYB.2020.2985081
25. Holland, J.H.: Adaptation in natural and artificial systems. MIT Press, Ann Arbor, MI (1975)
26. Horn, J., Nafpliotis, N., Goldberg, D.: A niched Pareto genetic algorithm for multiobjective optimization. In: Proceedings of the First IEEE Conference on Evolutionary Computation. IEEE World Congress on Computational Intelligence, vol. 1, pp. 82–87 (1994). https://doi.org/10.1109/ICEC.1994.350037
27. Hughes, E.J.: Evolutionary many-objective optimisation: Many once or one many? In: IEEE Congress on Evolutionary Computation (CEC-2005), pp. 222–227 (2005)

28. Inapakurthi, R.K., Mitra, K.: Optimal surrogate building using SVR for an industrial grinding process. Mater. Manuf. Processes **37**(15), 1701–1707 (2022). https://doi.org/10.1080/10426914.2022.2039699
29. Jaimes, A.L., Coello, C.A.C., Chakraborty, D.: Objective reduction using a feature selection technique. In: Genetic and Evolutionary Computation Conference (GECCO), pp. 673–680 (2008)
30. Jain, H., Deb, K.: An evolutionary many-objective optimization algorithm using reference-point based nondominated sorting approach, Part II: handling constraints and extending to an adaptive approach. IEEE Trans. Evol. Comput. **18**(4), 602–622 (2014). https://doi.org/10.1109/TEVC.2013.2281534
31. Khare, V., Yao, X., Deb, K.: Performance scaling of multi-objective evolutionary algorithms. In: Fonseca, C., Fleming, P., Zitzler, E., Thiele, L., Deb, K. (eds.) Evolutionary Multi-Criterion Optimization. Lecture Notes in Computer Science, vol. 2632, pp. 72–72. Springer, Berlin/Heidelberg (2003)
32. Knowles, J., Corne, D.: Quantifying the effects of objective space dimension in evolutionary multiobjective optimization. In: Obayashi, S., Deb, K., Poloni, C., Hiroyasu, T., Murata, T. (eds.) Evolutionary Multi-Criterion Optimization. Lecture Notes in Computer Science, vol. 4403, pp. 757–771. Springer, Berlin / Heidelberg (2007)
33. Koçer, H.G., Uymaz, S.A.: A novel local search method for LSGO with golden ratio and dynamic search step. Soft. Comput. **25**, 2115–2130 (2021). https://doi.org/10.1007/s00500-020-05284-x
34. Li, F., Gao, L., Shen, W., Cai, X., Huang, S.: A surrogate-assisted offspring generation method for expensive multi-objective optimization problems. In: 2020 IEEE Congress on Evolutionary Computation (CEC), pp. 1–8 (2020). https://doi.org/10.1109/CEC48606.2020.9185691
35. Li, K., Deb, K., Zhang, Q., Kwong, S.: An evolutionary many-objective optimization algorithm based on dominance and decomposition. IEEE Trans. Evol. Comput. **19**(5), 694–716 (2015). https://doi.org/10.1109/TEVC.2014.2373386
36. Li, K., Deb, K., Zhang, Q., Kwong, S.: An evolutionary many-objective optimization algorithm based on dominance and decomposition. IEEE Trans. Evol. Comput. **19**(5), 694–716 (2015)
37. Li, M., Yang, S., Liu, X.: Shift-based density estimation for Pareto-based algorithms in many-objective optimization. IEEE Trans. Evol. Comput. **18**(3), 348–365 (2014)
38. Liao, X., Li, Q., Zhang, W., Yang, X.: Multiobjective optimization for crash safety design of vehicle using stepwise regression model. Struct. Multidiscip. Optim. **35**, 561–569 (2008)
39. Lin, J., Liu, H., Peng, C.: The effect of feasible region on imbalanced problem in constrained multi-objective optimization. In: 2017 13th International Conference on Computational Intelligence and Security (CIS), pp. 82–86 (2017)
40. Liu, S., Lin, Q., Li, J., Tan, K.C.: A survey on learnable evolutionary algorithms for scalable multiobjective optimization. IEEE Trans. Evol. Comput. **27**(6), 1941–1961 (2023). https://doi.org/10.1109/TEVC.2023.3250350
41. Ma, Z., Wang, Y.: Evolutionary constrained multiobjective optimization: test suite construction and performance comparisons. IEEE Trans. Evol. Comput. **23**(6), 972–986 (2019). https://doi.org/10.1109/TEVC.2019.2896967
42. Mallipeddi, R., Lee, M.: Surrogate model assisted ensemble differential evolution algorithm. In: 2012 IEEE Congress on Evolutionary Computation, pp. 1–8 (2012). https://doi.org/10.1109/CEC.2012.6256479
43. Miettinen, K.: Nonlinear multiobjective optimization. Kluwer, Boston (1999)
44. Mittal, S., Saxena, D.K., Deb, K.: Learning-based multi-objective optimization through ANN–assisted online innovization. In: Proceedings of the 2020 Genetic and Evolutionary Computation Conference Companion, pp. 171–172. GECCO 20, Association for Computing Machinery, New York, NY (2020). https://doi.org/10.1145/3377929.3389925
45. Mittal, S., Saxena, D.K., Deb, K., Goodman, E.D.: A learning-based *innovized* progress operator for faster convergence in evolutionary multi-objective optimization. ACM Trans. Evol. Learn. Optim. **2**(1) (2021). https://doi.org/10.1145/3474059

46. Mittal, S., Saxena, D.K., Deb, K., Goodman, E.D.: Enhanced innovized progress operator for evolutionary multi- and many-objective optimization. IEEE Trans. Evol. Comput. **26**(5), 961–975 (2022). https://doi.org/10.1109/TEVC.2021.3131952
47. Mittal, S., Saxena, D.K., Deb, K., Goodman, E.D.: A unified innovized progress operator for performance enhancement in evolutionary multi- and many-objective optimization. IEEE Trans. Evol. Comput. p. 1 (2023). https://doi.org/10.1109/TEVC.2023.3321603
48. Miuhlenbein, H., Paaß, G.: From recombination of genes to the estimation of distributions I. binary parameters. In: Proceedings of the 4th International Conference on Parallel Problem Solving from Nature, pp. 178–187. London (1996)
49. Purshouse, R.C., Fleming, P.J.: On the evolutionary optimization of many conflicting objectives. IEEE Trans. Evol. Comput. **11**(6), 770–784 (2007)
50. Rachmawati, L., Srinivasan, D.: Multiobjective evolutionary algorithm with controllable focus on the knees of the Pareto front. IEEE Trans. Evol. Comput. **13**(4), 810–824 (2009)
51. Ren, Q., Luo, F., Ding, W., Lu, H.: An improved NSGAII algorithm based on site-directed mutagenesis method for multi-objective optimization. In: 2019 IEEE Symposium Series on Computational Intelligence (SSCI), pp. 176–181 (2019). https://doi.org/10.1109/SSCI44817.2019.9002847
52. Saxena, D.K., Duro, J.A., Tiwari, A., Deb, K., Zhang, Q.: Objective reduction in many-objective optimization: linear and nonlinear algorithms. IEEE Trans. Evol. Comput. **77**(1), 77–99 (2013)
53. Saxena, D., Deb, K.: Non-linear dimensionality reduction procedures for certain large-dimensional multi-objective optimization problems: employing correntropy and a novel maximum variance unfolding. In: Obayashi, S., Deb, K., Poloni, C., Hiroyasu, T., Murata, T. (eds.) Evolutionary Multi-Criterion Optimization. Lecture Notes in Computer Science, vol. 4403, pp. 772–787. Springer, Berlin/Heidelberg (2007)
54. Saxena, D.K., Kapoor, S.: On timing the nadir-point estimation and/or termination of reference-based multi- and many-objective evolutionary algorithms. In: Deb, K., Goodman, E., Coello Coello, C.A., Klamroth, K., Miettinen, K., Mostaghim, S., Reed, P. (eds.) Evolutionary Multi-Criterion Optimization, pp. 191–202. Springer, Cham (2019)
55. Saxena, D.K., Mittal, S., Deb, K., Goodman, E.D.: Machine Learning Assisted Evolutionary Multi- and Many- Objective Optimization. Springer, Singapore (2024). https:// doi.org/ 10.1007/978-981-99-2096-9
56. Saxena, D.K., Mittal, S., Kapoor, S., Deb, K.: A localized high-fidelity-dominance based many-objective evolutionary algorithm. IEEE Trans. Evol. Comput. 1 (2022). https://doi.org/10.1109/TEVC.2022.3188064
57. Schaffer, J.D.: Multiple objective optimization with vector evaluated genetic algorithms. Ph.D. thesis, Vanderbilt University (1984)
58. Schröder, B.S.W.: Ordered Sets: An Introduction. Birkhäuser, Boston (2003)
59. Singh, H.K., Isaacs, A., Ray, T.: A Pareto corner search evolutionary algorithm and dimensionality reduction in many-objective optimization problems. IEEE Trans. Evol. Comput. **15**(4), 539–556 (2011)
60. Sinha, A., Bedi, S., Deb, K.: Bilevel optimization based on kriging approximations of lower level optimal value function. In: 2018 IEEE Congress on Evolutionary Computation (CEC), pp. 1–8 (2018). https://doi.org/10.1109/CEC.2018.8477763
61. Srinivas, N., Deb, K.: Muiltiobjective optimization using nondominated sorting in genetic algorithms. Evol. Comput. **2**(3), 221–248 (1994). https://doi.org/10.1162/evco.1994.2.3.221
62. Veldhuizen, D.V., Lamont, G.B.: Multiobjective evolutionary algorithm research: a history and analysis. Technical Report TR-98-03. Department of Electrical and Computer Engineering, Air Force Institute of Technology, Dayton, OH (1998)
63. Wang, H., Jin, Y.: A random forest-assisted evolutionary algorithm for data-driven constrained multiobjective combinatorial optimization of trauma systems. IEEE Trans. Cybern. **50**(2), 536–549 (2020). https://doi.org/10.1109/TCYB.2018.2869674
64. Wang, R., Dong, N.J., Gong, D.W., Zhou, Z.B., Cheng, S., Wu, G.H., Wang, L.: PCA-assisted reproduction for continuous multi-objective optimization with complicated Pareto optimal set. Swarm Evol. Comput. **60**, 100795 (2021). https://doi.org/10.1016/j.swevo.2020.100795

65. Yuan, Y., Xu, H., Wang, B., Yao, X.: A new dominance relation-based evolutionary algorithm for many-objective optimization. IEEE Trans. Evol. Comput. **20**(1), 16–37 (2016). https://doi.org/10.1109/TEVC.2015.2420112
66. Zhang, Q., Li, H.: MOEA/D: a multiobjective evolutionary algorithm based on decomposition. IEEE Trans. Evol. Comput. **11**(6), 712–731 (2007). https://doi.org/10.1109/TEVC.2007.892759
67. Zhou, A., Jin, Y., Zhang, Q., Sendhoff, B., Tsang, E.: Combining model-based and genetics-based offspring generation for multi-objective optimization using a convergence criterion. In: 2006 IEEE International Conference on Evolutionary Computation, pp. 892–899 (2006). https://doi.org/10.1109/CEC.2006.1688406
68. Zhou, A., Zhang, Q., Jin, Y., Tsang, E., Okabe, T.: A model-based evolutionary algorithm for bi-objective optimization. In: 2005 IEEE Congress on Evolutionary Computation, vol. 3, pp. 2568–2575 (2005). https://doi.org/10.1109/CEC.2005.1555016
69. Zhou, Z., Wang, Z., Pang, T., Wei, J., Chen, Z.: A competition-cooperation evolutionary algorithm with bidirectional multi-population local search and local hypervolume-based strategy for multi-objective optimization. In: 2021 IEEE Congress on Evolutionary Computation (CEC), pp. 153–160 (2021). https://doi.org/10.1109/CEC45853.2021.9504689
70. Zitzler, E., Künzli, S.: Indicator-based selection in multiobjective search. In: Yao, X., Burke, E., Lozano, J., Smith, J., Merelo-Guervós, J., Bullinaria, J., Rowe, J., Tino, P., Kabán, A., Schwefel, H.P. (eds.) Parallel Problem Solving from Nature—PPSN VIII. Lecture Notes in Computer Science, vol. 3242, pp. 832–842. Springer, Berlin / Heidelberg (2004)
71. Zitzler, E., Thiele, L.: Multiobjective evolutionary algorithms: a comparative case study and the strength Pareto approach. IEEE Trans. Evol. Comput. **3**(4), 257–271 (1999). https://doi.org/10.1109/4235.797969
72. Zitzler, E., Deb, K., Thiele, L.: Comparison of multiobjective evolutionary algorithms: empirical results. Evol. Comput. **8**(2), 173–195 (2000). https://doi.org/10.1162/106365600568202

A Many-Objective Optimized Folded Cross Regression Model (FCRM) with Unspecified Targets: A Critical Analysis

Paramartha Dutta, Sourab Mandal, and Rishita Mondal

Abstract Optimizing complex systems, particularly those without clearly defined target variables, remains a challenge across many fields such as agriculture, where numerous interdependent factors influence outcomes. To address this, we present a hybrid method that combines linear regression with evolutionary many-objective optimization using the Non-dominated Sorting Genetic Algorithm III (NSGA-III). The proposed Folded Cross Regression Model (FCRM) identifies the Pareto front of optimal solutions by jointly minimizing prediction errors measured through both Root Mean Square Error (RMSE) and Mean Square Error (MSE) across all features. This formulation enables the discovery of robust regression configurations and highlights trade-offs between competing objectives without requiring predefined targets. We validate FCRM on multiple benchmark datasets, including Abalone, Wind Speed, and Bike Sharing, demonstrating that the model effectively reveals Pareto-optimal relationships and structural dependencies within data-driven systems. FCRM thus provides a general framework for analyzing domains where target variables are ambiguous, context-dependent, or absent.

Keywords Many-objective optimization · NSGA-III · Genetic algorithms · Linear regression · Pareto front · Folded Cross Regression Model · Complex systems

1 Introduction

In many real-world systems, particularly in domains such as agriculture, environment, and medicine, optimizing outcomes is difficult when there is no clearly defined target variable. For instance, in agriculture, crop yield depends on a complex interaction of factors such as soil composition, weather conditions, irrigation, and fertilizer use. In such cases, standard supervised learning methods are inadequate because they require a specific, known outcome for model training. It is important to clarify that this

P. Dutta · S. Mandal (✉) · R. Mondal
Department of Computer and System Sciences, Visva-Bharati, Santiniketan, India
e-mail: smvb.rs@gmail.com

S. Mukhopadhyay et al. (eds.), *Contemporary Advancement in Evolutionary Multi-objective Optimization*, Genetic and Evolutionary Computation,
https://doi.org/10.1007/978-981-95-7908-2_3

limitation does not imply that conventional machine learning methods are ineffective. Rather, these methods are inherently designed for settings where the dependent variable is clearly defined. In target-ambiguous domains such as agriculture, environmental monitoring, or energy systems, multiple interdependent variables may serve as potential outputs, making it difficult to specify a single predictive objective without bias. For example, in crop modeling, yield, nutrient content, and water use efficiency may each be valid but competing outcomes, depending on the context of optimization. Folded Cross Regression Model (FCRM) is therefore introduced not as a replacement for supervised learning, but as a complementary exploratory framework that can analyze such inter-feature dependencies when the target itself is uncertain or evolving. To address this limitation, we propose a regression framework that functions without a predefined target by treating each feature in the dataset as a potential target in turn. The approach integrates linear regression with the Non-dominated Sorting Genetic Algorithm III (NSGA-III) [2] to perform many-objective optimization. The objective is to minimize prediction errors, measured using both Root Mean Square Error (RMSE) and Mean Square Error (MSE), across all features in the dataset. This process identifies structurally dependent variables and yields regression configurations that are robust under multiple criteria. The method produces a Pareto front [3] of non-dominated solutions, each representing a distinct trade-off in prediction error across objectives. Unlike conventional regression, which optimizes a single target variable, the proposed FCRM evaluates all features simultaneously, making it suitable for domains where the concept of a fixed target is ill-defined or context-dependent. The effectiveness of FCRM is demonstrated using several benchmark datasets, including Abalone, Wind Speed, and Bike Sharing, showing that the model can reveal interpretable, target-agnostic patterns and relationships. Although agricultural data serve as a motivating example, the framework is applicable to a wide range of complex systems involving interdependent variables. The remainder of this paper is organized as follows. Section 1 presents the motivation and background of the study. Section 2 discusses preliminary concepts, related work, and the identified research gap. Section 3 describes the FCRM model, its structure, algorithm, and interpretation. Section 4 provides the datasets and experimental results. Section 5 summarizes the contributions and future research directions, and Sect. 6 lists links to code, datasets, and supplementary materials.

2 Background

This section introduces the preliminary terminologies, reviews related works, identifies existing research gaps, outlines the motivation behind the study, and highlights the salient contributions of the proposed approach.

2.1 Preliminary Terminologies

In multi-objective optimization, each candidate solution is represented by a vector of decision variables that defines a point in the *decision space*, while the corresponding objective values define a point in the *objective space*. The goal is to discover solutions that balance conflicting objectives effectively [1, 8–10].

Definition 1 (Multi-objective and Many-Objective Optimization) Multi-objective optimization involves simultaneously optimizing two or more conflicting objective functions. Formally, it seeks a solution vector $\mathbf{x} \in \mathcal{D}$ that minimizes or maximizes a set of objective functions:

$$\mathbf{f}(\mathbf{x}) = [f_1(x), f_2(x), \ldots, f_m(x)],$$

where $m \geq 2$ and each $f_i(x)$ is an objective function. A solution is considered optimal if it lies on the *Pareto front* of non-dominated solutions. When the number of objectives becomes large ($m \geq 4$), the problem is referred to as a *many-objective optimization* problem, which introduces additional challenges in maintaining both convergence and diversity among solutions [1, 8–10].

Definition 2 (Pareto Optimality) A solution is Pareto-optimal if no other solution improves one objective without degrading at least one other. The collection of all such non-dominated solutions forms the *Pareto front*, which represents the trade-off surface among competing objectives [1, 8–10].

Definition 3 (Fitness Vector) In the context of the proposed FCRM, a *fitness vector* represents the set of prediction errors obtained when each feature in the dataset is treated, in turn, as the target variable. Each element of the vector corresponds to the error associated with one such configuration, forming the basis for many-objective optimization using NSGA-III [2].

2.2 Literature Review

Optimization involving multiple, often conflicting, objectives has long been a central challenge in computational intelligence. A major advancement in this area was the Non-dominated Sorting Genetic Algorithm II (NSGA-II) proposed by Deb et al. [3], which introduced a fast non-dominated sorting approach and elitism to maintain convergence quality. Although NSGA-II performs effectively for problems with a small number of objectives, its scalability decreases as dimensionality increases, primarily due to reduced selection pressure and difficulty in preserving population diversity. To overcome these limitations, Deb and Jain proposed NSGA-III [2], which extends NSGA-II by employing reference-point-based selection to maintain a well-distributed set of Pareto-optimal solutions in high-dimensional objective spaces. This approach has proven particularly suitable for many-objective optimization, forming the foundation of the strategy adopted in the present study. Parallel to these developments, the integration of regression modeling with evolutionary algorithms has received increasing attention. Such hybrid methods aim to combine the predictive capability of regression with the exploratory power of evolutionary search. For instance, Devi et al. [4] proposed a genetic algorithm–assisted multiple linear regression model for drug efficacy prediction, while Wang et al. [14] introduced a regression-driven optimization framework capable of autonomously configuring multi-objective algorithms under constraints. Similarly, the MOEA/D algorithm proposed by Zhang and Li [16] decomposes a multi-objective problem into a set of scalar sub-problems, enabling simultaneous optimization of interconnected objectives. These works collectively demonstrate the utility of regression-guided optimization for complex problem spaces, yet they generally presume a fixed and well-defined target variable. However, in many real-world domains such as environmental modeling, agriculture, and engineering design, defining a single response variable is often impractical or arbitrary. Existing optimization frameworks rarely address such situations where the target variable itself is uncertain or evolving. This gap motivates the development of the FCRM, which unifies regression analysis and NSGA-III–based many-objective optimization into a single framework. By cyclically treating each feature as a potential target, FCRM explores inter-feature dependencies without requiring a predefined objective, generating Pareto-optimal regression vectors that reveal structural relationships across all variables.

2.3 Gap Identification and Motivation

Although many-objective optimization methods such as NSGA-III [2] and regression-guided evolutionary models [4, 14] have shown success across a range of applications, they typically rely on the assumption of a clearly defined target variable. This assumption simplifies model formulation but limits applicability in complex, real-world systems where no single measurable objective exists. Domains like agriculture, environmental systems, or engineering design often involve multiple interrelated factors, such as cost, performance, or sustainability, that defy singular optimization. In such settings, traditional methods struggle due to their dependence on fixed objective structures. This work addresses that gap by proposing an optimization strategy that does not require a predefined target. Instead, it evaluates each feature in the dataset as a potential target, enabling a many-objective exploration of possible trade-offs. This motivates the development of the FCRM, which integrates linear regression and NSGA-III to identify robust, Pareto-optimal configurations in target-agnostic systems.

2.4 Salient Contributions of the Work

The key contributions of this work are summarized below.

1. Proposed the FCRM, a target-unspecified optimization framework for systems without a fixed output variable.
2. Combined multiple linear regression with NSGA-III to enable many-objective optimization in a target-agnostic setting.
3. Introduced a folded evaluation strategy in which each variable is cyclically treated as the temporary target to capture inter-feature dependencies.
4. Formulated a fitness vector of prediction errors (RMSE and MSE) to guide the search toward Pareto-optimal regression vectors.
5. Employed NSGA-III to identify stable trade-offs among reconstruction errors across all candidate target variables. The regression coefficients and intercept jointly represent a solution in the decision space, while the fitness vector defines its corresponding position in the objective space.
6. Validated the model on multiple benchmark datasets to demonstrate robustness and general applicability.
7. Generated Pareto-optimal fronts that describe the trade-off landscape of RMSE and MSE objectives for each dataset. Feature ranking based on Pareto solutions is identified as a potential direction for future work, as it extends beyond the current study's focus on Pareto-front analysis.

3 FCRM: Folded Cross Regression Model with Unspecified Target

The proposed FCRM addresses optimization problems where the target variable is either undefined or varies contextually across applications. Unlike conventional models that rely on a fixed output, FCRM systematically evaluates each feature as a potential dependent variable. The model combines multiple linear regression for prediction and NSGA-III for many-objective optimization, forming a hybrid framework suited for complex, multi-variable systems.

3.1 Overview of the Approach

Figure 1b presents the core framework of the Folded Cross Regression Model (FCRM). For a dataset D containing n numeric variables, each feature $x_i \in D$ is sequentially treated as a temporary target, while the remaining variables act as predictors. A regression vector $\boldsymbol{\alpha} = [\alpha_0, \alpha_1, \ldots, \alpha_n]$ models the joint relationships among variables. Predicted values are compared to the actual targets to

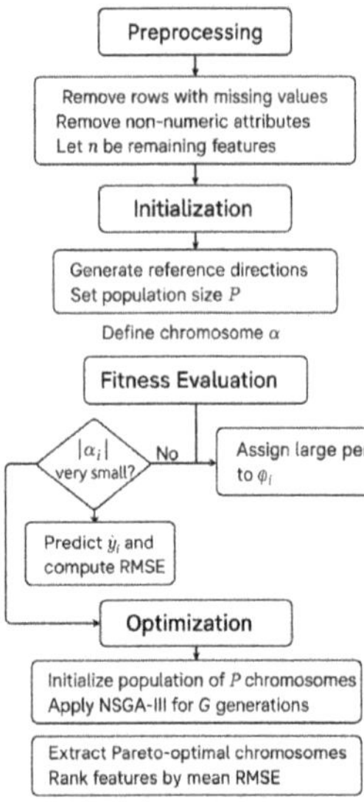

(a) Flow Diagram for FCRM

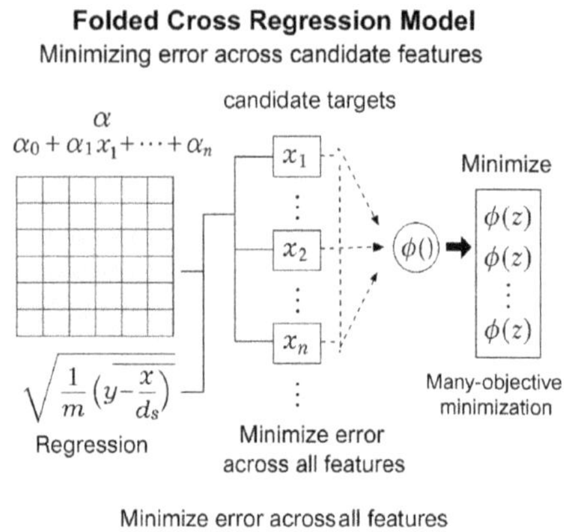

(b) Folded Cross Regression Model (FCRM)

Fig. 1 Illustration of the Folded Cross Regression Model: **a** workflow steps based on Algorithm 1; **b** conceptual schematic

obtain error metrics such as RMSE or MSE. This process generates a fitness vector $\boldsymbol{\phi}(\boldsymbol{\alpha}) = [\phi_1, \phi_2, \ldots, \phi_n]$, where each component represents the prediction error for a specific feature when used as the dependent variable. Minimizing this vector forms a many-objective optimization problem. The NSGA-III algorithm is employed to evolve a population of regression vectors toward the Pareto front, identifying trade-offs among competing feature-wise objectives.

3.2 Chromosome Representation

Each solution candidate is represented as a vector of regression coefficients:

$$\boldsymbol{\alpha} = [\alpha_0, \alpha_1, \ldots, \alpha_n]$$

where α_0 is the intercept and α_i is the coefficient corresponding to the ith feature. To eliminate arbitrary scaling effects, each chromosome is normalized using its Euclidean norm during evaluation.

3.3 Fitness Evaluation

The fitness of each chromosome is assessed by folding over all features so that every feature is in turn considered as the target. Prior to evaluation, the dataset is standardized to zero mean and unit variance, ensuring comparability of objectives. For each fold, if the coefficient α_i is smaller than a predefined threshold τ, the corresponding objective ϕ_i is penalized to prevent numerical instability. Otherwise, the predicted value of x_i is computed from the regression equation:

$$\hat{x}_i = -\frac{\alpha_0}{\alpha_i} - \sum_{j \neq i} \frac{\alpha_j}{\alpha_i} x_j$$

Residuals are obtained as $\epsilon_{pi} = x_{pi} - \hat{x}_{pi}$ for each sample p. The mean squared error (MSE) and root mean squared error (RMSE) are then calculated as:

$$\text{MSE}_i = \frac{1}{m} \sum_{p=1}^{m} \epsilon_{pi}^2, \qquad \text{RMSE}_i = \sqrt{\text{MSE}_i}$$

The vector $\boldsymbol{\phi} = [\phi_1, \ldots, \phi_n]$ is formed using either RMSE or MSE, depending on the optimization objective.

3.4 Optimization with NSGA-III

The population of chromosomes is evolved using the NSGA-III algorithm, which employs Pareto dominance ranking and reference-direction diversity preservation in the objective space. During each generation, selection, simulated binary crossover, and polynomial mutation operators are applied to explore the solution space. The process continues for a predefined number of generations G, progressively refining the population toward a set of non-dominated solutions representing balanced trade-offs across all objectives.

3.5 Output and Interpretation

The final outcome of the FCRM is a collection of Pareto-optimal regression vectors that jointly minimize the prediction errors across all candidate target features. Each chromosome on the Pareto front defines a potential compromise among objectives, offering insight into variable interdependence within the dataset. Post-processing of the Pareto-optimal set allows computation of per-feature statistics such as the mean or variance of RMSE values, which can be used to assess relative predictability. Features that consistently exhibit lower average error are interpreted as being more explainable by the remaining variables. These results provide a data-driven understanding of mutual relationships among features without requiring a pre-specified target variable.

3.6 Algorithmic Description

Algorithm 1 outlines the operational steps of the FCRM. The flow of the procedure is illustrated in Fig. 1a, while Fig. 1 summarizes the model at a conceptual level. The process begins with preprocessing, where missing values and non-numeric attributes are removed, and all numeric variables are standardized. Initialization involves generating reference directions and setting up a population of chromosomes, each representing a normalized regression vector. During fitness evaluation, for each chromosome and feature x_i, the algorithm checks for near-zero coefficients, predicts the corresponding target, and computes the associated RMSE or MSE. The NSGA-III procedure iteratively evolves the population for G generations, converging toward a Pareto-optimal set of regression vectors. The resulting front is then analyzed to interpret feature relationships and assess the consistency of predictions across folds.

4 Experimental Evaluation

This section presents the experimental validation of the proposed FCRM. The model is evaluated on publicly available benchmark datasets. Performance is analyzed using standard error metrics and visualizations of the resulting Pareto fronts to assess robustness and optimization quality.

4.1 Datasets Used

Experiments were conducted on the datasets given in Table 1 to evaluate the generalizability and applicability of FCRM. It is worth mentioning that all datasets were preprocessed by normalizing continuous features and removing categorical or missing data. No predefined target variable was set during optimization.

4.2 Experimental Setup

All experiments were performed on Google Colab using a Python 3 runtime hosted on the Google Compute Engine backend. The runtime environment provided approximately 12.7 GB of RAM and 108 GB of disk space. The proposed FCRM was implemented in `Python` using the `pymoo` optimization library for NSGA-III,

Algorithm 1 Folded Cross Regression Model (FCRM) with unspecified targets

Require: Raw dataset D with m instances and n numeric attributes; number of generations G; near-zero threshold $\tau > 0$

Ensure: Pareto-optimal regression vectors and ranked features

Preprocessing

Remove rows with missing values

Keep only numeric attributes; let n be the remaining feature count

Standardize each feature to zero-mean, unit-variance (z-score) so objectives are comparable

Augment data with an intercept column of ones

Decision Encoding and Initialization

A chromosome is $\boldsymbol{\alpha} = [\alpha_0, \alpha_1, \ldots, \alpha_n]$ (intercept first)

To remove scale ambiguity, normalize $\boldsymbol{\alpha}$ at evaluation time, e.g., $\boldsymbol{\alpha} \leftarrow \boldsymbol{\alpha}/\|\boldsymbol{\alpha}\|_2$

Generate reference directions for n objectives (e.g., Das–Dennis); set population size P accordingly

Initialize a population of P chromosomes (e.g., Latin Hypercube or random)

Folded Fitness Evaluation (per chromosome)

for $i = 1$ **to** n **do**

  {treat feature x_i as the target in fold i}

  if $|\alpha_i| \le \tau$ **then**

    Set $\phi_i \leftarrow \Phi_{\max}$ {large penalty to avoid division instability}

    continue

  end if

  Solve the implicit hyperplane $\alpha_0 + \sum_{k=1}^{n} \alpha_k x_k = 0$ for x_i:

$$\beta_{i0} = -\frac{\alpha_0}{\alpha_i}, \qquad \beta_{ij} = -\frac{\alpha_j}{\alpha_i} \quad (j \neq i)$$

  Predict: $\hat{x}_{pi} = \beta_{i0} + \sum_{j \neq i} \beta_{ij} x_{pj}$ for $p = 1, \ldots, m$

  Residuals: $\epsilon_{pi} = x_{pi} - \hat{x}_{pi}$

  Compute per-fold errors:

$$\mathrm{MSE}_i = \frac{1}{m} \sum_{p=1}^{m} \epsilon_{pi}^2, \qquad \mathrm{RMSE}_i = \sqrt{\mathrm{MSE}_i}$$

  Set objective value $\phi_i \leftarrow \mathrm{RMSE}_i$ {default; see remark below for using MSE as well}

end for

Multiobjective Optimization (NSGA-III)

for $g = 1$ **to** G **do**

  Evaluate $\boldsymbol{\phi} = [\phi_1, \ldots, \phi_n]$ for each chromosome

  Apply NSGA-III (selection, SBX crossover, polynomial mutation, reference-direction niching)

end for

Outputs

Extract the final Pareto front of chromosomes (w.r.t. $\boldsymbol{\phi}$)

Table 1 Details of datasets used for evaluation

Dataset	Known target variable(s)	Instances (P)	Features (n)
Abalone [11]	Rings	4177	8
Bike sharing [5]	Count (Cnt)	17,389	13
Concrete comprehensive strength [15]	Concrete compressive strength	1030	8
Energy efficiency [13]	$Y1$ and $Y2$	768	8
Friedman [7]	Y	40,768	11
Parkinsons telemonitoring [12]	motor_UPDRS and total_UPDRS	5875	19
Wind speed [6]	Wind	6574	8

together with standard scientific computing libraries such as `NumPy`, `Pandas`, and `scikit-learn`. The same NSGA-III configuration was applied to all datasets:

- Population size (P): 120
- Number of generations (G): 120
- Crossover probability (p_c): 1.0
- Mutation probability (p_m): $\frac{1}{n+1}$, where n is the number of features
- Distribution index for crossover (η_c): 20
- Distribution index for mutation (η_m): 20
- Reference-direction generation: Das–Dennis scheme with $n_{\text{partitions}} = 1$
- Near-zero coefficient threshold (τ): 10^{-6}
- Penalty for unstable folds ($\Phi_{\max}$): 10^6.

Each run produced $2n$ objective functions representing the RMSE and MSE values for each feature when treated as the temporary target. The initial population was generated with random coefficients uniformly sampled from the range $[-5, 5]$ and evolved through simulated binary crossover and polynomial mutation. All experiments were executed with a fixed random seed (`42`) to ensure reproducibility of the evolutionary process. Reported results correspond to the outcome of this controlled run for each dataset.

4.3 Results and Analysis

The proposed FCRM was evaluated on eight datasets representing diverse domains, including environmental, behavioral, energy, biomedical, and synthetic data. The datasets comprise Abalone, Bike Sharing (Day), Bike Sharing (Hour), Concrete, Energy Efficiency, Fried, Parkinsons, and Wind data, as summarized in Table 1. Each dataset was processed using the configuration described in Sect. 4.2, and optimization was performed using NSGA-III to obtain Pareto-optimal solutions for the joint minimization of RMSE and MSE across all features. For every dataset, the FCRM generated a collection of non-dominated solutions representing different trade-offs between reconstruction errors of the individual features when each was cyclically treated as a target. The resulting Pareto fronts capture the balance between minimizing RMSE and MSE simultaneously, reflecting the multi-objective nature of the folded regression process. Each Pareto-optimal point corresponds to a distinct regression vector that performs well across multiple target substitutions, without the need for specifying a fixed dependent variable.

The parallel coordinate plots serve as a visual summary of the multi-objective optimization results, allowing the comparative evaluation of folded regression targets and highlighting the trade-offs among them. The Pareto fronts obtained from the proposed FCRM for all benchmark datasets are presented in Figs. 2, 3, 4, 5, 6, 7, 8, and 9. Each plot illustrates the normalized RMSE values across the folded regression targets, with every line representing one Pareto-optimal solution produced by NSGA-III. The visualization for the Abalone dataset (Fig. 2) shows moderate variation among folds, suggesting comparable predictive strength across multiple variables. In the Bike Sharing (Day) and (Hour) datasets (Figs. 3 and 4), several lines intersect, indicating trade-offs between daily and hourly predictive objectives. The Concrete dataset (Fig. 5) displays more distinct separation among folds, revealing that certain target variables dominate the optimization outcomes. For the Energy Efficiency dataset (Fig. 6), the lines remain relatively parallel, suggesting consistent

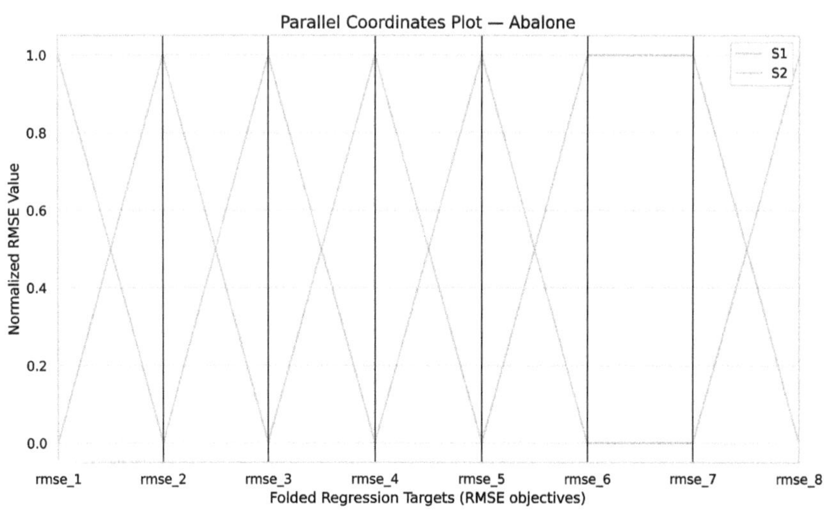

Fig. 2 Parallel coordinates of normalized RMSE objectives for the Abalone dataset. Moderate variation among folds indicates comparable predictive strength across variables

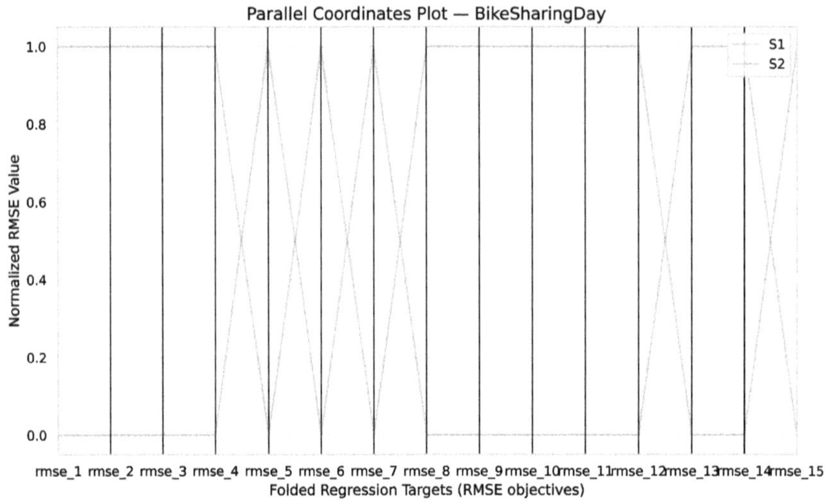

Fig. 3 Parallel coordinates of normalized RMSE objectives for the Bike Sharing (Day) dataset. Intersections among lines indicate trade-offs between daily predictive objectives

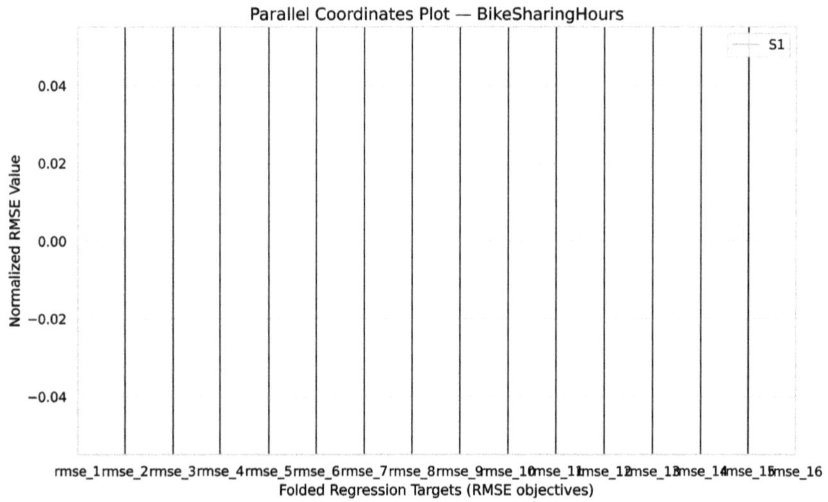

Fig. 4 Parallel coordinates of normalized RMSE objectives for the Bike Sharing (Hour) dataset. Crossing patterns reflect trade-offs across hourly predictive objectives

performance across all folds and highlighting the model's stability. The Fried dataset (Fig. 7) exhibits higher dispersion, reflecting a wider diversity of Pareto-optimal solutions. In the Parkinsons dataset (Fig. 8), the spread of lines implies variable-level sensitivity and non-uniform error distribution among objectives. Finally, the Wind

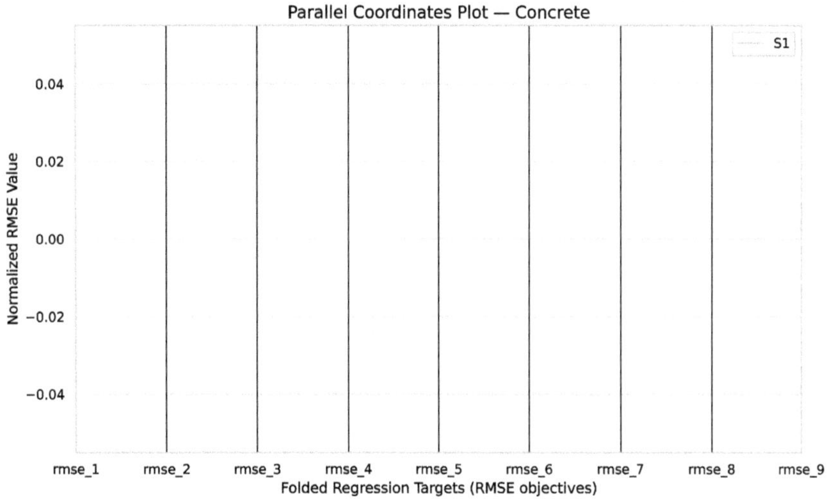

Fig. 5 Parallel coordinates of normalized RMSE objectives for the Concrete dataset. Distinct separation among folds suggests certain target variables dominate the optimization outcomes

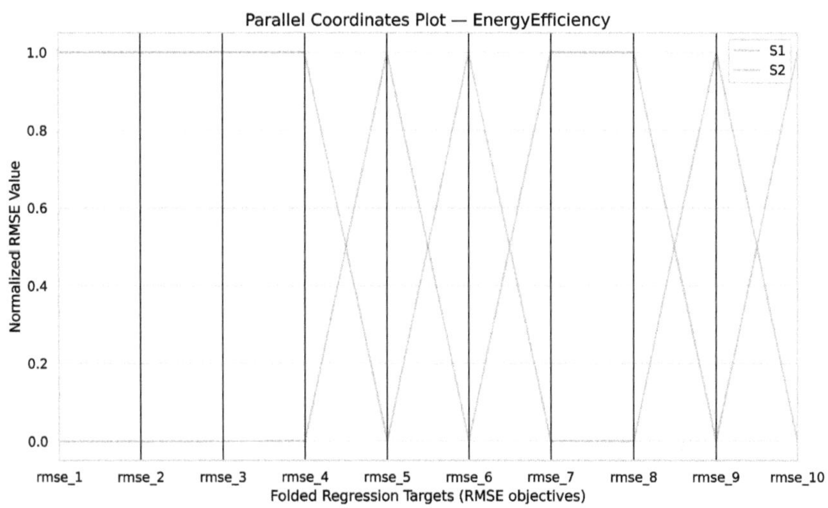

Fig. 6 Parallel coordinates of normalized RMSE objectives for the Energy Efficiency dataset. Lines remain relatively parallel, indicating consistent performance across folds

dataset (Fig. 9) demonstrates balanced patterns with only minor trade-offs, showing that the model effectively captures interdependencies without sacrificing overall performance. These observations collectively confirm that the proposed framework can

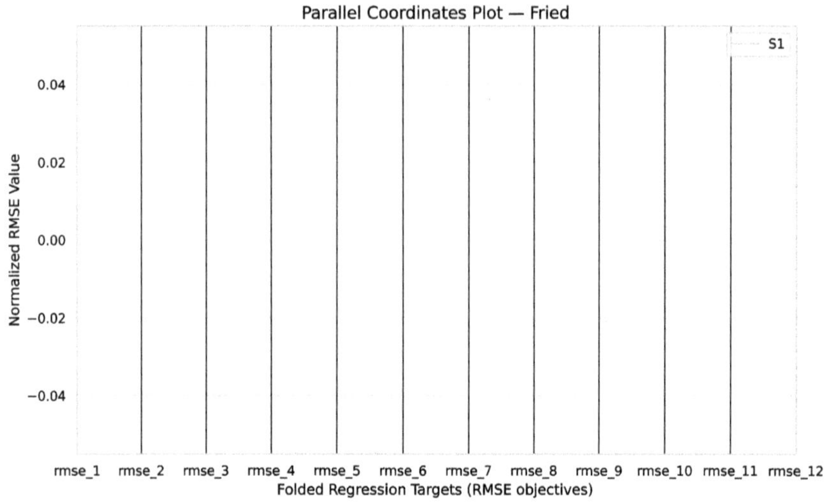

Fig. 7 Parallel coordinates of normalized RMSE objectives for the Fried dataset. Higher dispersion reflects a wider diversity of Pareto-optimal solutions

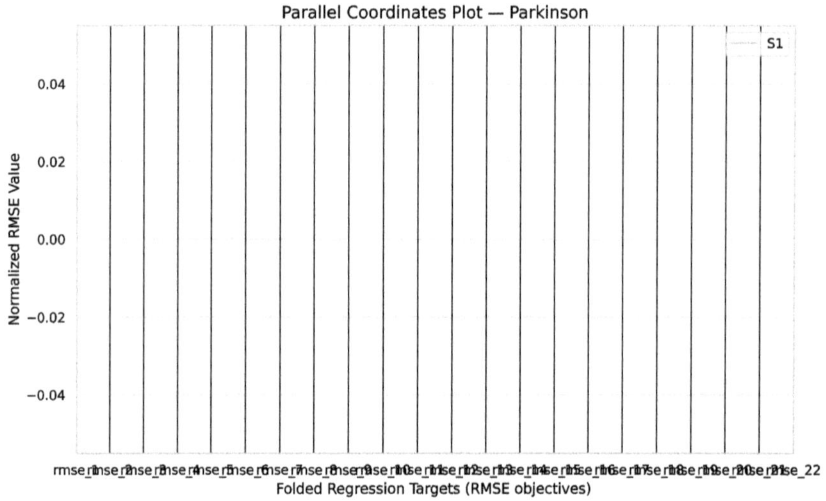

Fig. 8 Parallel coordinates of normalized RMSE objectives for the Parkinson dataset. The spread of lines implies variable-level sensitivity and non-uniform error distribution among objectives

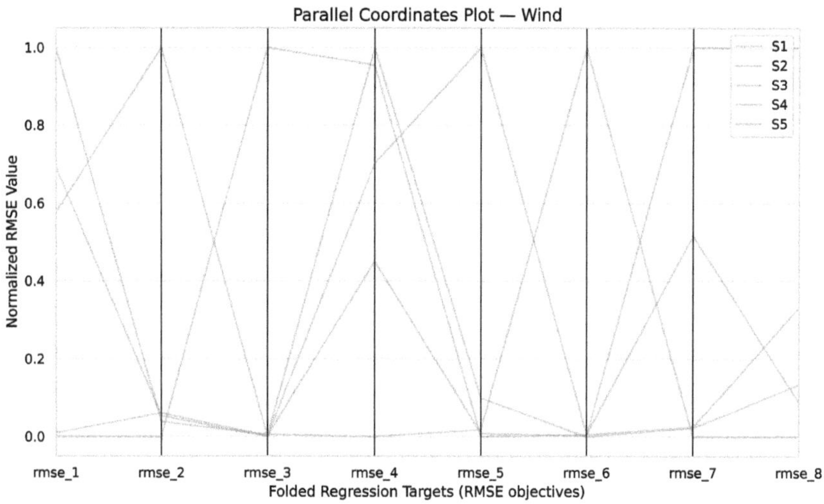

Fig. 9 Parallel coordinates of normalized RMSE objectives for the Wind dataset. Balanced patterns with minor trade-offs show captured interdependencies without loss of overall performance

adapt to diverse data structures while maintaining a well-distributed set of solutions across objectives.

The complete Pareto-optimal solution sets for each dataset, including the corresponding RMSE and MSE values, are provided as supplementary CSV files accompanying this study.

5 Conclusion and Future Scope

In this work, we presented the FCRM, a many-objective optimization framework designed to analyze regression problems without relying on a predefined target variable. By cyclically treating each feature as a potential target, FCRM formulates a multiobjective search problem that simultaneously minimizes reconstruction errors across all features. The proposed approach, implemented using NSGA-III, enables a target-agnostic exploration of inter-feature dependencies and generates a diverse set of Pareto-optimal regression vectors. Experiments conducted on eight benchmark datasets demonstrated that the model effectively identifies meaningful structural relationships among variables, producing smooth and consistent Pareto fronts for RMSE and MSE objectives. To ensure transparency and reproducibility, the complete Pareto-optimal solution sets for all datasets have been made available as supplementary CSV files.

Future research will explore ranking strategies derived from Pareto-optimal sets, as well as the integration of sparsity-inducing regularization and feature selection mechanisms to enhance interpretability and model compactness. Further extensions may include nonlinear variants using kernel or neural formulations and adaptations of FCRM for classification or causal analysis tasks. Applying the model to large-scale, high-dimensional domains such as genomics, remote sensing, and social systems may also reveal complex interdependencies that are difficult to capture through conventional supervised methods.

6 Supplementary Materials

All the codes, used datasets, and results, which include the CSV files related to the Pareto optimal solutions, are provided as supplementary material via Zenodo. The link for the same is https://doi.org/10.5281/zenodo.17453225.

Acknowledgements The authors are grateful to the Department of Computer and System Sciences, Visva-Bharati, Santiniketan, WB, India for providing us with a conducive environment to conduct the research smoothly.

Disclosure of Interests The authors declare that they have no relevant financial or non-financial interests that could have influenced the content of this article.

References

1. Deb, K.: Multi-objective optimisation using evolutionary algorithms: an introduction. In: Multi-objective Evolutionary Optimisation for Product Design and Manufacturing, pp. 3–34. Springer (2011)
2. Deb, K., Jain, H.: An evolutionary many-objective optimization algorithm using reference-point-based nondominated sorting approach, part I: solving problems with box constraints. IEEE Trans. Evol. Comput. **18**(4), 577–601 (2013)
3. Deb, K., Pratap, A., Agarwal, S., Meyarivan, T.: A fast and elitist multiobjective genetic algorithm: NSGA-II. IEEE Trans. Evol. Comput. **6**(2), 182–197 (2002)
4. Devi, A.S., Hertono, G.F., Sarwinda, D., Siswantining, T.: Combining of genetic algorithm and multiple linear regression in breast cancer's drug design. In: 2018 2nd International Conference on Biomedical Engineering (IBIOMED), pp. 121–125 (2018). https://doi.org/10.1109/IBIOMED.2018.8533960
5. Fanaee-T, H., Gama, J.: Event labeling combining ensemble detectors and background knowledge. Prog. Artif. Intell. **2**(2), 113–127 (2014). https://doi.org/10.1007/s13748-013-0040-3
6. fedesoriano: Wind speed prediction dataset. https://www.kaggle.com/datasets/fedesoriano/wind-speed-prediction-dataset (2022). Accessed 29 Apr 2026
7. Friedman, J.H.: Multivariate adaptive regression splines. Ann. Stat. **19**(1), 1–67 (1991)
8. Mandal, S., Dutta, P.: Multi-objective non-overlapping front generation: a pivot-based deterministic non-dominated sorting approach. In: International Conference on Pattern Recognition and Machine Intelligence, pp. 559–567. Springer (2023)
9. Mandal, S., Dutta, P., Ghosh, A.: Decomposition into multi-objective fronts: a pivot-based deterministic non-dominated sorting approach. In: International Conference on Human-Centric Smart Computing, pp. 55–66. Springer (2023)

10. Mukherjee, A., Mandal, S., Dutta, P.: Generating multi-objective fronts from streamed data using nested list. In: International Conference on Pattern Recognition, pp. 128–143. Springer (2025)
11. Nash, W., Sellers, T., Talbot, S., Cawthorn, A., Ford, W.: Abalone Dataset. UCI Machine Learning Repository (1995). https://doi.org/10.24432/C55C7W
12. Tsanas, A., Little, M.: Parkinsons Telemonitoring. UCI Machine Learning Repository (2009). https://doi.org/10.24432/C5ZS3N
13. Tsanas, A., Xifara, A.: Energy Efficiency. UCI Machine Learning Repository (2012). https://doi.org/10.24432/C51307
14. Wang, Y., Sun, X., Zhang, Y., Gong, D., Hu, H., Zuo, M.: Linear regression-based autonomous intelligent optimization for constrained multi-objective problems. IEEE Trans. Artif. Intell. **1**(2), 121–132 (2020). https://doi.org/10.1109/TAI.2020.3031793
15. Yeh, I.C.: Concrete Compressive Strength. UCI Machine Learning Repository (1998). https://doi.org/10.24432/C5PK67
16. Zhang, Q., Li, H.: MOEA/D: a multiobjective evolutionary algorithm based on decomposition. IEEE Trans. Evol. Comput. **11**(6), 712–731 (2007). https://doi.org/10.1109/TEVC.2007.892759

Emerging Techniques for Evolutionary Single- and Multi-objective Optimization with Application to Humanoid Robot Gait Generation

Pushpendra Gupta, Dilip Kumar Pratihar, and Kalyanmoy Deb

Abstract This chapter presents emerging techniques in evolutionary computation that address critical challenges such as fair algorithm comparison, dynamic performance evaluation, and knee-based multi-objective optimization. These techniques are applied to a real-world problem of gait generation for bipedal locomotion. For single-objective optimization, a novel statistical comparison methodology is proposed that employs multiple independent runs to gain insight into algorithm consistency and reliability. The Particle swarm optimization (PSO) algorithm demonstrates superior convergence characteristics and reduced performance variability compared to a genetic algorithm (GA) across different initial conditions. In multi-objective optimization, the study introduces running performance metrics for a dynamic assessment without requiring knowledge of the true Pareto-optimal front (PF). The running Inverted Generational Distance Plus (IGD+) indicator reveals that elitist non-dominated sorting GA (NSGA-II) is superior to multi-objective PSO (MOPSO) in handling constraints. Furthermore, knee-based optimization techniques focusing on PF regions are discussed to reduce decision-maker burden. The angle-based knee detection method is validated on a DO2DK benchmark problem generation wise and later applied to humanoid robot gait optimization, demonstrating its effectiveness

P. Gupta
Department of Data Science and Artificial Intelligence, BITS Pilani Digital, Birla Institute of Technology and Science, Bengaluru, Karnataka, India
e-mail: pushpendra.gupta@bitspilani-digital.edu.in

D. K. Pratihar (✉)
Department of Mechanical Engineering, Indian Institute of Technology Kharagpur, Kharagpur, West Bengal, India
e-mail: dkpra@mech.iitkgp.ac.in

K. Deb
Department of Electrical and Computer Engineering, Michigan State University, East Lansing, MI, USA
e-mail: kdeb@egr.msu.edu

S. Mukhopadhyay et al. (eds.), *Contemporary Advancement in Evolutionary Multi-objective Optimization*, Genetic and Evolutionary Computation,
https://doi.org/10.1007/978-981-95-7908-2_4

in identifying critical trade-off regions. The single-support phase exhibits one distinct knee region while the double-support phase reveals two, reflecting the underlying behavior of bipedal locomotion. These applications demonstrate the usefulness of evolutionary single- and multi-objective optimization algorithms in solving a complex real-world optimization problem with multiple objectives.

Keywords Genetic algorithm · Particle swarm optimization · Evolutionary computation · Humanoid robotics · Knee solutions

1 Introduction

Evolutionary computation (EC) has emerged as a powerful tool for solving complex optimization problems. These nature-inspired algorithms have shown remarkable success across diverse fields including engineering design [1, 2], quantum computing [3], robotics [4–6], machine learning [7], and bioinformatics [8]. They have demonstrated their niche particularly in handling multi-modal, highly constrained, noisy, and non-linear optimization problems. Traditional gradient-based methods often get trapped in local optima for such problems.

In humanoid robotics, evolutionary algorithms have become essential for addressing bipedal gait generation [9, 10]. They provide robust solutions where conventional approaches fail to handle the complex dynamics and constraints of human-like walking. Recent applications of evolutionary multi-objective optimization (EMO) algorithms in bipedal robotics include trajectory optimization for navigation [11], fuzzy adaptive control for trajectory tracking [12], and stability-energy trade-off analysis [13], demonstrating their effectiveness in generating efficient and stable gaits.

Selecting an appropriate optimization algorithm remains a fundamental challenge. The No Free Lunch theorem [14] states that no single algorithm can perform best on all problems. This principle applies to both single- and multi-objective scenarios. For single-objective problems, the challenge becomes evident when different algorithms converge to different solutions with dependency on the initial population. Genetic algorithm (GA) [15] and particle swarm optimization (PSO) [16] algorithms may reach the same or slightly different optimum, but despite their similarity at the core principle, they employ different search strategies. The GA relies on crossover and mutation operations inspired by natural evolution. The PSO uses swarm intelligence principles based on collective behavior. It is clear from the above discussion that each algorithm has its own niche of problem types. Thus, it becomes an important task to compare these different EC methods for a specific problem class, such as the robot gait generation problem.

Algorithm selection becomes more complex for multi-objective optimization problems (MOP). Unlike single-objective problems, multi-objective optimization (MOO) generates a set of Pareto-optimal (PO) solutions exhibiting trade-offs among conflicting objectives. An EMO algorithm finds this PO set in a single run, contrary to single-objective methods that produce one solution per run. This fundamental difference makes algorithm comparison challenging. Traditional performance indicators like generational distance [17] (GD, GD+) and inverted generational distance [18] (IGD, IGD+) require knowledge of the true Pareto front (PF), which is rarely available in real-world problems [5]. This chapter addresses this limitation by proposing reference set construction methods to approximate the true PF. Moreover, dynamic algorithm behavior, including convergence patterns and stagnation phases, remains hidden when comparing only final-generation results, yet significantly affects overall performance. This chapter usages running performance metrics to reveal such intermediate behaviors during optimization.

There is another decision-making challenge associated with a MOP. The sheer number of PO solutions an EMO algorithm produce can be overwhelming to the decision-makers (DMs). Practitioners need additional ways to focus on specific regions of the PF for an easier decision-making purpose. "Knee" solutions [19], if exist in a problem, offer superior trade-offs and deserve special attention. By identifying these regions, one can reduce hundreds of solutions to a manageable set while preserving critical information. Thus, finding knee solutions directly without resorting to finding the entire PF can be a computationally effective strategy.

Humanoid robots are anthropomorphic systems designed to mimic human movement and behavior. Their human-like form allows them to work in environments built for people. They can navigate factory floors, hospital corridors, and other human spaces effectively. This makes them valuable for various applications, from hazardous material handling to healthcare assistance. As these robots find wider use, optimizing their walking patterns becomes important. Two key objectives often are in conflict in bipedal locomotion. First, we want to minimize power consumption for longer operational time. Second, we need to maximize stability for a safe and reliable movement. These objectives work against each other: stable gaits often require more energy. This trade-off demands sophisticated MOO approaches. We must find optimal solutions that meet operational needs while maintaining the physical limits.

This chapter addresses these challenges by presenting emerging techniques in EC field through the case study of a humanoid robot gait generation problem, which demonstrates both theoretical advancements and practical applications. The research employs a systematic approach, formulating the gait optimization problem in both single- and multi-objective contexts to showcase different algorithmic capabilities and comparison methodologies. For single-objective optimization focusing on power consumption minimization, the study introduces novel statistical comparison frameworks for evaluating GA and PSO performance across multiple independent runs, addressing the stochastic nature of evolutionary algorithms. The multi-objective formulation extends the analysis to simultaneously consider power consumption and stability, employing the NSGA-II and MOPSO with advanced performance evaluation techniques including running performance metrics [20] for a dynamic algorithm assessment and knee-based solution [21] identification for an enhanced decision-making support.

The methodological contributions presented herein discuss emerging techniques in EC by introducing systematic frameworks for algorithm comparison, novel performance indicators that operate without reference sets, and practical techniques for reducing cognitive burden in multi-objective decision-making. These contributions, demonstrated through the challenging problem of gait optimization for a 25 degrees-of-freedom (DOF) NAO humanoid robot [22], exhibit broad applicability across complex engineering domains where multiple conflicting objectives must be balanced under stringent constraints.

This chapter is organized as follows: Sect. 2 establishes the theoretical foundations of single- and multi-objective optimization, covering PF characteristics and emerging comparison methodologies. Section 3 presents the mathematical formulation of the humanoid robot gait generation problem, including objective functions and constraints for both gait phases. Section 4 applies GA and PSO to single-objective optimization, introducing novel statistical comparison frameworks for fair algorithm evaluation. Section 5 extends the analysis to MOO using NSGA-II and MOPSO, incorporating running performance metrics and knee-based solution identification techniques. Finally, Sect. 6 synthesizes key insights and outlines future research directions.

2 Emerging Techniques in Evolutionary Computation

The field of EC has witnessed significant advancements from traditional optimization methods to sophisticated techniques capable of handling complex, real-world problems. This evolution has been particularly pronounced in the development of algorithms for single and MOO scenarios. The emergence of new techniques addresses fundamental limitations of classical approaches while introducing novel methodologies for algorithm comparison and solution selection.

2.1 Single- and Multi-objective Optimization Fundamentals

Optimization problems in engineering and scientific applications involve finding the best combination of decision variables that optimizes certain performance criteria while satisfying imposed constraints. A general optimization problem can be mathematically formulated as:

$$\underset{\mathbf{x}\in\mathbb{R}^N}{\text{minimize}} \quad \{f_1(x), f_2(x), \ldots, f_M(x)\}, \tag{1a}$$

subject to

$$g_j(x) \leq 0, \quad j = 1, \ldots, J, \tag{1b}$$

$$h_k(x) = 0, \quad k = 1, \ldots, K, \tag{1c}$$

$$x_i^{\mathrm{L}} \leq x_i \leq x_i^{\mathrm{U}}, \quad i = 1, \ldots, N, \tag{1d}$$

where $x = [x_1, x_2, \ldots, x_N]^{\mathrm{T}}$ represents the vector of N decision variables, $f_1(x), f_2(x), \ldots, f_M(x)$ denote M objective functions to be optimized, $g_j(x) \leq 0$ represent J inequality constraints, $h_k(x) = 0$ denote K equality constraints, and x_i^{L} and x_i^{U} define the lower and upper bounds for the ith decision variable, respectively.

The nature of the optimization problem fundamentally changes based on the number of objectives M. When $M = 1$, the problem falls into the category of single-objective optimization, characterized by the existence of usually one, but possibly multiple local or global, optimal solutions. The optimization process seeks the best possible combination of decision variables that minimizes (or maximizes) the sole objective function while satisfying all constraints. Traditional optimization algorithms, both gradient-based and EC algorithms, can effectively handle such problems by converging to a local or global optimum through iterative improvement of the objective function value. Due to the population approach and their stochastic search process, EC algorithms have a better chance of finding the global optimal solutions.

In contrast, when M is two or three, the problem belongs to the MOO category, presenting fundamentally different characteristics. The defining feature of MOPs lies in the conflicting nature of the objectives, that is, an improvement in one objective typically results in deterioration of at least one other objective. This conflict eliminates the possibility of finding a single solution that simultaneously optimizes all objectives. Instead, MOO generates a set of trade-off solutions, known as PO solutions, where no solution can be improved in any objective without degrading at least one other objective.

The concept of Pareto-dominance provides the mathematical framework for comparing solutions in multi-objective space. A solution x^* is said to dominate another solution x if and only if $f_i(x^*) \leq f_i(x)$ for all $i \in \{1, 2, \ldots, M\}$ and $f_j(x^*) < f_j(x)$ for at least one $j \in \{1, 2, \ldots, M\}$. Solutions that are not dominated by any other feasible solution constitute the PO set, and their mapping in the objective space forms the PF. Based on dominance relationships, solutions can be ranked where Rank 1 represents non-dominated solutions, Rank 2 represents solutions dominated only by Rank 1, and so on. Figure 1a illustrates this ranking hierarchy, where Rank 1 solutions (red circles) represent the PO solutions closest to the true PF, while dominated solutions of Rank 2 (blue squares), Rank 3 (black diamonds), and Rank 4 (green triangles) can each be improved in at least one objective without degrading others.

This fundamental distinction between single and MOO necessitates different algorithmic approaches, performance evaluation metrics, and decision-making strategies. While single-objective optimization focuses on convergence to the global optimum, MOO must balance convergence toward the true PF while maintaining diversity across the front to provide comprehensive trade-off information to the DMs.

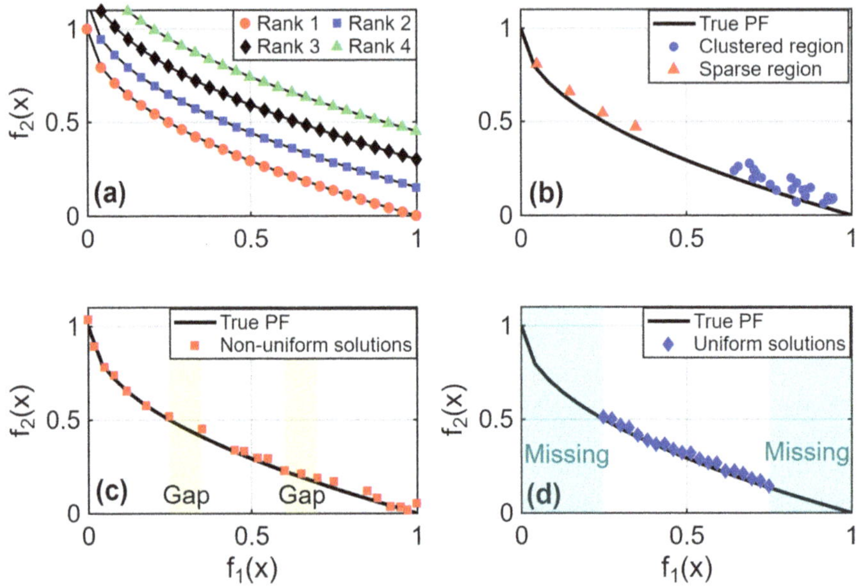

Fig. 1 Illustration of PF quality characteristics: **a** convergence with multiple ranks, **b** good cardinality but imbalanced distribution, **c** good spread but poor uniformity with irregular spacing, **d** good uniformity but poor spread with missing extreme solutions

2.2 *Characteristics of Pareto Front*

EMO algorithms generate numerous solutions on a PF, making direct comparison between different algorithm outputs challenging. A large number of solutions does not automatically indicate superior algorithm performance. Four key characteristics provide a systematic evaluation framework.

Convergence measures how closely solutions approximate the true PF. Figure 1a illustrates this through multiple fronts with different ranks, where Rank 1 represents best convergence. Diversity ensures solutions distribute well across the entire PF. Poor diversity creates uneven distribution despite adequate solution count, as shown in Fig. 1b where clustering occurs in lower regions while upper regions remain sparse. Uniformity refers to even spacing between consecutive solutions along the PF. Figure 1c shows good spread with poor uniformity through irregular spacing and gaps, while Fig. 1d demonstrates good uniformity with poor spread by missing extreme solutions. Cardinality represents the total number of non-dominated solutions, though distribution quality matters more than sheer quantity.

Understanding these characteristics enables fair algorithm comparison and informed selection for practical applications. An algorithm producing fewer but well-distributed solutions often provides more value than one generating numerous clustered solutions. The specific type of solutions generated with particular uniformity patterns or gaps remains problem-dependent, but comparing two PFs from different algorithms consistently requires evaluation of convergence quality, spread coverage, uniformity, and cardinality. The ideal PFs exhibit high convergence to the true optimal front, comprehensive diversity with complete coverage, good uniformity with even spacing, and appropriate cardinality balancing computational cost with decision-making needs. These evaluation criteria become essential when selecting algorithms for specific problem classes, such as the robot gait generation problem addressed in this chapter.

3 Humanoid Robotics Problem Definition

This section presents the established mathematical formulation of the humanoid robot gait generation problem for the 25-DOF NAO robot, developed and validated in our previous works [4–6]. This well-characterized problem serves as the application for demonstrating the emerging optimization techniques that constitute the primary contribution of this chapter: statistical comparison methodologies for single-objective algorithms (Sect. 4), running performance metrics for MOO (Sect. 5), and knee-based solution identification for enhanced decision support (Sect. 5.3). The bipedal locomotion of humanoid robots comprises two fundamental phases that alternate cyclically to produce stable walking patterns as shown in Fig. 2. To better understand the details of these phases, Fig. 3 shows various aspects of bipedal locomotion: a schematic view during single support phase (SSP) with stride length from x_i to x_f, hip height Z_H, and maximum swing height $S_\mathrm{h}^\mathrm{max}$, the definition of zero moment point (ZMP) and dynamic balance margin (DBM), a schematic view during double support phase (DSP), ZMP's maximum and minimum values depending on support leg location, NAO robot foot dimensions, and minimum X_DBM and Y_DBM values during DSP [6].

During the SSP, the robot maintains balance on one leg while the contralateral leg executes a swing motion to advance forward. This phase represents the

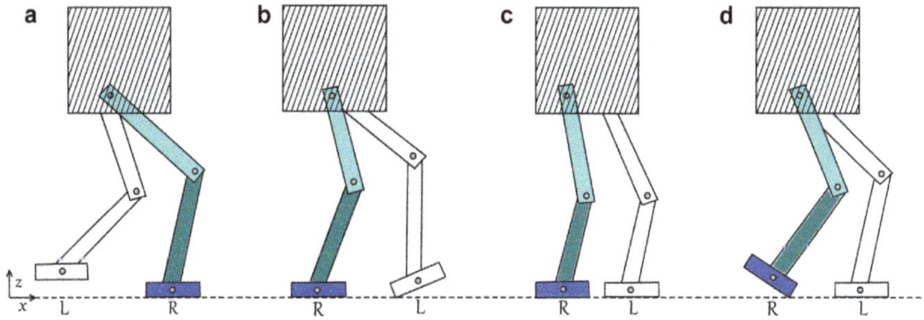

Fig. 2 A schematic view showing four phases of a locomotion cycle of biped robot: **a** single support phase, **b** contact phase, **c** double support phase, **d** take-off phase (where R and L represent right and left feet, respectively)

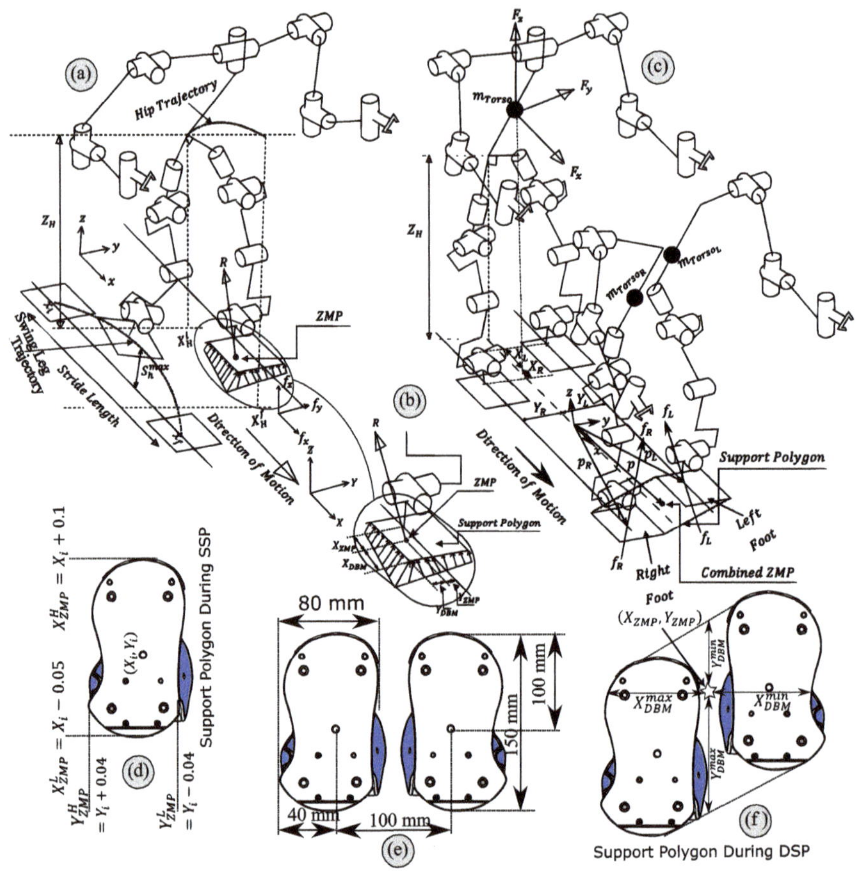

Fig. 3 Detailed illustration of bipedal locomotion mechanics: **a** schematic view during SSP showing swing leg trajectory, hip height (Z_{H}), and maximum swing height ($S_{\mathrm{h}}^{\max}$), **b** definition of zero moment point and dynamic balance margin showing how the resultant force must remain within the support polygon, **c** schematic view during DSP with combined ZMP calculation, **d** ZMP bounds based on support leg location, **e** NAO robot foot dimensions, **f** minimum X_{DBM} and Y_{DBM} calculations during DSP [6]

most dynamically challenging aspect of bipedal locomotion, as the robot operates as an open kinematic chain extending from the support foot through the torso. The swing leg trajectory is characterized by its movement from initial position x_{i} to final position x_{f}, achieving a maximum swing height $S_{\mathrm{h}}^{\max}$ at the midpoint of the swing trajectory. Throughout this phase, the hip follows a three-dimensional trajectory $[X_{\mathrm{H}}(t), Y_{\mathrm{H}}(t), Z_{\mathrm{H}}(t)]^{\mathrm{T}}$, transitioning from $[x_{\mathrm{i}} + x_{\mathrm{f}}/4, Y_{\mathrm{H}}, Z_{\mathrm{h}}]^{\mathrm{T}}$ to $[x_{\mathrm{i}} + 3x_{\mathrm{f}}/4, Y_{\mathrm{H}}, Z_{\mathrm{h}}]^{\mathrm{T}}$ to ensure forward progression while maintaining dynamic stability.

The DSP occurs when both feet establish contact with the ground, creating a closed kinematic structure that provides enhanced stability through an enlarged support polygon. During this transitional phase, the robot's center of mass shifts from the previous support leg to the next, with the hip executing a lateral movement. The torso mass distribution between the legs is determined by their relative sagittal and lateral distances from the respective feet.

The focus of this chapter is on the optimization methodologies and their comparative analysis rather than detailed gait characteristics. Comprehensive discussions about kinematic analysis, joint trajectory profiles, and Webots simulations from the optimized solutions for single objective optimization (SOO) have been presented in our previous work [4]. The current formulation builds upon these validated models to demonstrate emerging techniques in EC, with the humanoid robot serving as a representative complex optimization problem exhibiting multiple conflicting objectives, nonlinear constraints, and high-dimensional decision spaces.

3.1 Objective Functions

The gait optimization problem addresses two conflicting objectives that represent the fundamental trade-offs in bipedal locomotion. The first objective $f_1(x)$ minimizes the average power consumption across all active joints during the gait cycle:

$$f_1(x) = \sum_{j=1}^{n} \frac{1}{T} \int_{0}^{T} \left(\left|\tau_j \dot{\theta}_j\right| + K\tau_j^2\right) \mathrm{d}t, \tag{2}$$

where n denotes the number of active joints, T represents the gait cycle duration, τ_j and $\dot{\theta}_j$ are the torque and angular velocity of the jth joint respectively, and $K = 0.025$ accounts for heat dissipation in the actuators. The second objective $f_2(x)$ maximizes the dynamic stability margin through minimization of its reciprocal:

$$f_2(x) = \frac{1}{X_{\mathrm{DBM}}}, \tag{3}$$

where X_{DBM} represents the DBM [23] in the sagittal direction. The DBM is computed as the minimum distance between the ZMP and the support polygon boundary during the gait cycle [24]. The ZMP is the point on the ground where the net moment of ground reaction forces equals zero. The ZMP must remain within the support polygon for dynamically stable bipedal locomotion.

3.2 Constraint Functions and Decision Variables

The optimization formulation incorporates multiple constraints to ensure physically realizable and stable locomotion patterns. The constraint functions are systematically organized to address joint limitations, stability requirements, operational smoothness, and coordinated movement patterns. The joint angle constraints $g_1(x)$ and $g_2(x)$ enforce that all joint rotations remain within their mechanical limits, preventing hardware damage and ensuring feasible configurations throughout the gait cycle:

$$g_1(x) \equiv \theta^j - \theta^j_{\text{min}} \geq 0, \tag{4}$$

$$g_2(x) \equiv \theta^j_{\text{max}} - \theta^j \geq 0, \tag{5}$$

where θ^j_{min} and θ^j_{max} represent the minimum and maximum allowable joint angles for the jth joint, respectively, based on the NAO robot's kinematic specifications [25]. The minimum balance constraints $g_3(x)$ and $g_4(x)$ maintain a minimum DBM of 0.001 m in both sagittal and lateral directions, ensuring that the ZMP remains within the support polygon with adequate safety margins:

$$g_3(x) \equiv \min\left(X^i_{\text{DBM}}\right) - 0.001\ \text{m} \geq 0, \tag{6}$$

$$g_4(x) \equiv \min\left(Y^i_{\text{DBM}}\right) - 0.001\ \text{m} \geq 0, \tag{7}$$

where X^i_{DBM} and Y^i_{DBM} represent the instantaneous DBM values in the sagittal and lateral directions during the ith time interval of the gait cycle. The average DBM constraints $g_5(x)$ and $g_6(x)$ impose phase-specific requirements on the lateral and sagittal stability margins:

$$g_5(x) \equiv Y^{\text{Avg}}_{\text{DBM}} - {}^{\text{min}}Y^{\text{Avg}}_{\text{DBM}} \geq 0, \tag{8}$$

$$g_6(x) \equiv X^{\text{Avg}}_{\text{DBM}} - {}^{\text{min}}X^{\text{Avg}}_{\text{DBM}} \geq 0, \tag{9}$$

where $Y^{\text{Avg}}_{\text{DBM}}$ and $X^{\text{Avg}}_{\text{DBM}}$ represent the average DBM values throughout the gait cycle, and ${}^{\text{min}}Y^{\text{Avg}}_{\text{DBM}}$ and ${}^{\text{min}}X^{\text{Avg}}_{\text{DBM}}$ define the minimum required average stability margins. The torque fluctuation constraint $g_7(x)$ ensures smooth actuator operation by limiting the maximum torque variation across time intervals:

$$g_7(x) \equiv \Delta\tau_{\text{max}} - \Delta\tau^{\text{max}}_{ij} \geq 0, \tag{10}$$

where $\Delta\tau^{\text{max}}_{ij}$ represents the maximum torque fluctuation for the jth joint in the ith time interval, and $\Delta\tau_{\text{max}}$ defines the allowable torque variation threshold. The arm movement constraints $g_8(x)$ and $g_9(x)$ enforce proper coordination between upper and lower body movements, enhancing overall dynamic balance through synchronized limb motion:

$$g_8(x) \equiv \theta^{\text{final}}_{\text{LShoulderPitch}} - \theta^{\text{initial}}_{\text{LShoulderPitch}} \geq 0, \tag{11}$$

$$g_9(x) \equiv \theta^{\text{final}}_{\text{LElbowRoll}} - \theta^{\text{initial}}_{\text{LElbowRoll}} \geq 0, \tag{12}$$

ensuring that the left arm moves in coordination with the right leg swing motion, while the right arm moves oppositely to maintain balance. The decision variable vector x encompasses parameters controlling hip trajectory, velocity profiles, swing characteristics, and temporal aspects. For the SSP, nine decision variables are employed in SOO: hip height (x_1), initial and final velocities associated with hip height (x_2, x_3), initial and final sagittal velocities (x_4, x_5), initial and final lateral velocities (x_6, x_7), maximum swing height (x_8), and SSP duration (x_9). In MOO, additional arm motion parameters are included, expanding the decision space to thirteen variables with left arm ShoulderPitch angles (x_{10}, x_{11}) and ElbowRoll angles (x_{12}, x_{13}) for coordinated upper body movement.

3.3 Scope and Limitations

The current formulation focuses on kinematic and dynamic constraints essential for stable bipedal locomotion on flat, uniform surfaces. Environmental factors such as terrain variations, surface friction coefficients, and complex contact dynamics are not explicitly modeled in this study. The optimization framework assumes ideal contact conditions with known friction properties typical of laboratory environments. Future extensions could incorporate terrain-adaptive constraints and friction-dependent stability criteria for enhanced real-world applicability, though such additions would significantly increase problem complexity and computational requirements.

4 Single-Objective Optimization

This section presents the SOO approach for minimizing power consumption during the NAO robot's gait cycle, focusing specifically on the SSP. A similar formulation can be applied to the DSP with appropriate modifications to the constraint values and decision variables. The optimization problem for SSP is formulated as:

$$
\begin{aligned}
&\underset{\mathbf{x}\in\mathcal{X}\subseteq\mathbb{R}^9}{\text{minimize}} \quad \mathcal{P}_{\text{avg}} \text{ (refer to Eq. (2))},\\
&\text{subject to}\\
&\qquad g_1(x) \text{ through } g_7(x) \text{ given in Eqs. (4)–(10)},\\
&\qquad x_i^{\mathrm{L}} \le x_i \le x_i^{\mathrm{U}},
\end{aligned}
$$

with

$$
\begin{aligned}
x^{\mathrm{L}} &= [0.25, -0.05, -0.05, 0.001, 0.001, -0.2, -0.2, 0.015, 0.4]^{\mathrm{T}},\\
x^{\mathrm{U}} &= [0.31, 0.05, 0.05, 0.2, 0.2, 0.2, 0.2, 0.030, 4]^{\mathrm{T}}.
\end{aligned}
$$

The constraint parameters taken for ${}^{\min}Y_{\mathrm{DBM}}^{\mathrm{Avg}}$, ${}^{\min}X_{\mathrm{DBM}}^{\mathrm{Avg}}$, and $\Delta\tau_{ij}^{\max}$ are 0.02 m, 0.05 m, and 1.2 Nm, respectively. The nine decision variables encompass hip height, initial and final velocities for hip height changes, sagittal and lateral velocity components, maximum swing height, and SSP duration.

4.1 Genetic Algorithm and Particle Swarm Optimization Parameters

GA employs crossover and mutation operations inspired by natural selection to evolve a population of candidate solutions toward optimal regions, while PSO utilizes collective intelligence principles where particles move through the search space based on their own experience and swarm knowledge. Both algorithms are well-established metaheuristics capable of handling complex, nonlinear optimization problems with multiple constraints.

A real-coded GA is employed with simulated binary crossover (SBX) having distribution index $\eta_c = 2$ and crossover probability $P_c = 0.9$. Polynomial mutation operates with distribution index $\eta_m = 20$ and mutation probability $P_m = 0.1$. The population size is maintained at 50 individuals. PSO is configured with inertia weight of 1.0, cognitive parameter of 1.5, social parameter of 2.0, and swarm size of 50 particles. Both algorithms execute for 200 generations to ensure sufficient convergence analysis while maintaining reasonable computational requirements.

The selection of GA and PSO for single-objective comparison represents two fundamentally different evolutionary paradigms: genetic operations (crossover and mutation) versus swarm intelligence (particle velocity and position updates). Similarly, elitist non-dominated sorting GA (NSGA-II) and multi-objective PSO (MOPSO) are chosen for multi-objective analysis as representative algorithms of their respective paradigms with proven track records in constrained optimization. While newer algorithms such as MOEA/D, CMA-ES variants, and Differential Evolution could provide additional insights, the primary contribution of this chapter lies in the comparison methodology rather than exhaustive algorithm benchmarking. The proposed statistical comparison framework and running performance metrics are broadly applicable to any evolutionary or metaheuristic optimizer. Extension to contemporary algorithms represents a natural direction for future investigation.

4.2 Proposed Statistical Comparison

The evaluation of metaheuristic algorithms presents significant challenges due to their inherent stochastic nature, where different runs may yield varying results even with identical parameter settings. Traditional single-run comparisons fail to capture algorithm consistency and reliability, necessitating comprehensive statistical evaluation methodologies that account for performance variability across multiple independent runs.

Figure 4 illustrates the complete statistical comparison framework developed for fair algorithm evaluation. Two methodologies can be employed: first, generating eleven different sets of initial populations and running each algorithm once per set, or second, using one initial population set and running each algorithm eleven times with the same starting conditions. Both approaches collect comprehensive performance data at each generation, recording average fitness of the population and best fitness value found. For each generation across all eleven runs, six critical statistical measures are calculated: minimum, average, and maximum values for both population average fitness and best fitness metrics. Since the objective is minimization, the

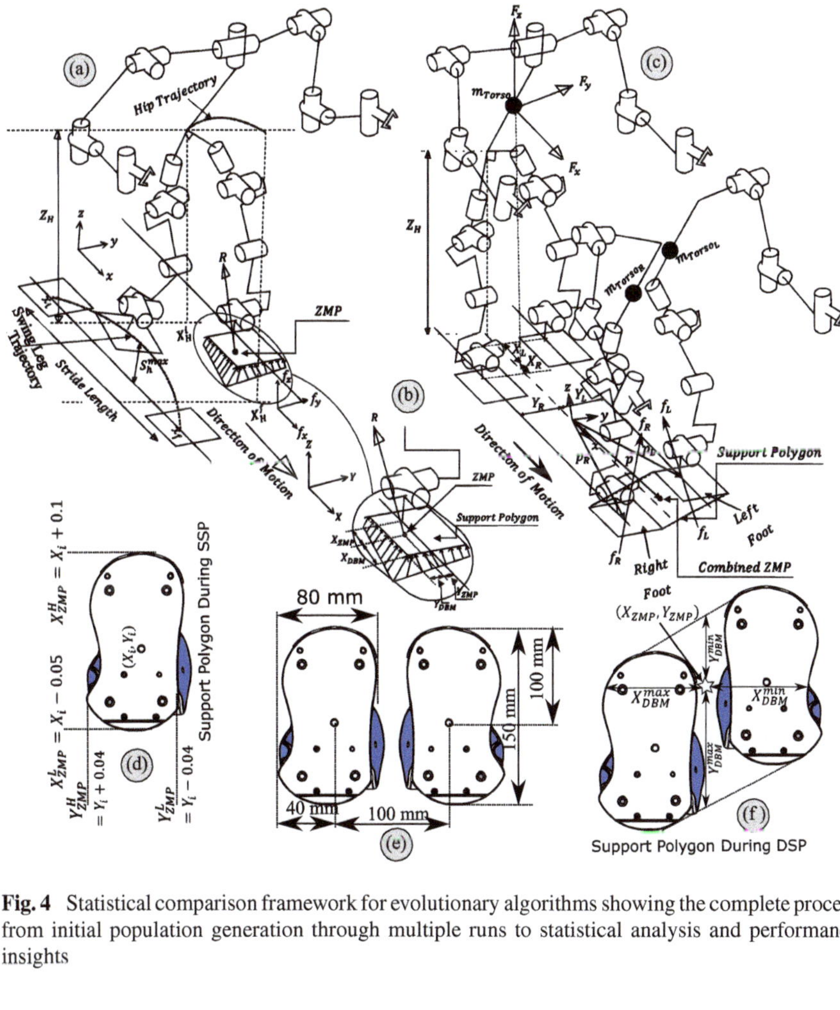

Fig. 4 Statistical comparison framework for evolutionary algorithms showing the complete process from initial population generation through multiple runs to statistical analysis and performance insights

minimum values represent the best performance, providing insights into algorithm capabilities, consistency, and reliability across different optimization scenarios.

4.3 Results and Discussion

Our previous work [4] employed a comprehensive comparison methodology using eleven independent runs for each algorithm, analyzing median performance to provide unbiased algorithm evaluation. The comparison utilized a single initial population set run eleven times for both GA and PSO, with analysis focused on minimum fitness values achieved in each generation across all runs. Results demonstrated PSO's superior performance in achieving lower power consumption with better convergence characteristics and solution consistency compared to GA during SSP optimization.

A comparable analysis using average fitness values across generations provides additional insights into algorithm behavior, as shown in Fig. 5. The error bars represent minimum and maximum average values across all eleven runs, indicating algorithm consistency where smaller error bars suggest greater reliability. PSO (refer to Fig. 5a) demonstrates superior performance with significantly smaller error bars in later generations, indicating more consistent results across multiple runs. This analysis reveals run-to-run variability even when algorithms start with identical initial populations, with PSO exhibiting higher reliability and consistency compared to GA (refer to Fig. 5b).

Figure 6 illustrates the convergence behavior of the best-performing runs for both algorithms over the first 50 generations, providing clear evidence of PSO's superior early-stage performance. PSO exhibits steeper descent in initial generations,

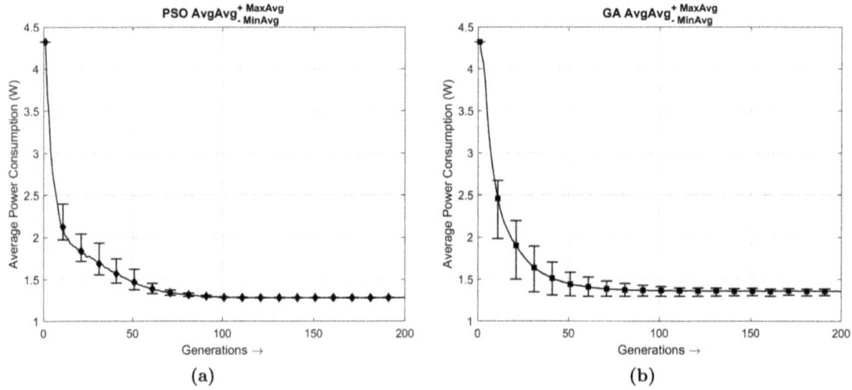

Fig. 5 Comparison of MaxAvg (maximum of average values) and MinAvg (minimum of average values) values with error bars with respect to the average of average objective values in each generation, obtained by **a** PSO and **b** GA during the 11 runs in SSP using the same set of initial population for all runs

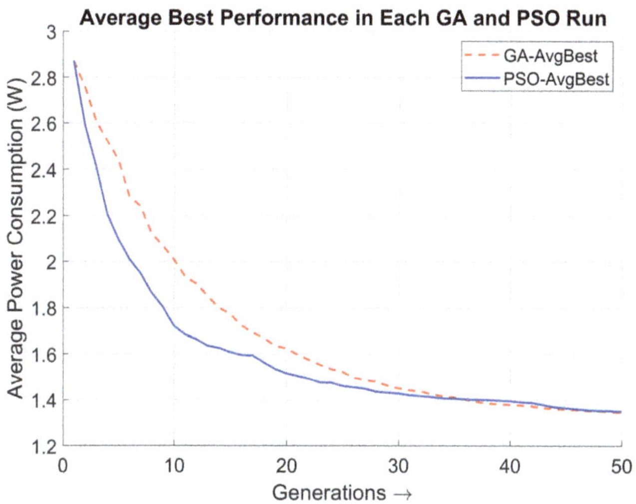

Fig. 6 Comparison of average of the population-best objective value between GA and PSO over 50 generations during the single support phase

demonstrating rapid identification and exploitation of promising search regions. By approximately the 10th generation, PSO achieves significantly lower average power consumption compared to GA, indicating superior exploration efficiency and faster convergence to optimal solutions through its effective information sharing mechanism and particle memory capabilities.

The statistical analysis using different initial populations for eleven independent runs reveals fundamental algorithmic characteristics through comprehensive performance profiling. Figure 7a shows GA's statistical performance with relatively wide performance variations, particularly in early generations, indicating sensitivity to initial population configurations. Conversely, Fig. 7b demonstrates PSO's markedly different behavior characterized by rapid early convergence and tighter performance bands across all runs. The narrow range between maximum and minimum values for PSO suggests excellent algorithm consistency with minimal sensitivity to initial configurations, while GA's wider performance gaps indicate greater variability and dependence on starting conditions, confirming PSO's superior reliability for practical robotics applications.

5 Multi-objective Optimization

This section presents the MOO formulation for simultaneous power minimization and stability maximization in humanoid robot gait generation. The analysis employs NSGA-II [26] and MOPSO [27] algorithms to generate PO solutions representing diverse trade-offs between conflicting objectives, incorporating coordinated arm movements for enhanced dynamic balance.

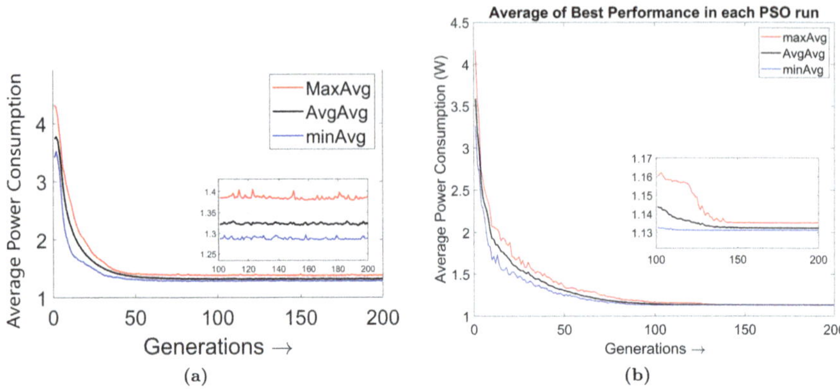

Fig. 7 Statistical performance analysis of **a** GA and **b** PSO during single support phase optimization showing maximum, average, and minimum objective values across 11 runs over 200 generations

$$\begin{aligned} &\underset{x\in\mathcal{X}\subseteq\mathbb{R}^{13}}{\text{minimize}} \quad \{f_1(x), f_2(x)\}, \\ &\text{subject to} \\ &\qquad g_1(x) \text{ through } g_9(x) \text{ given in Eqs. (4)–(12)}, \\ &\qquad x_i^{\mathrm{L}} \le x_i \le x_i^{\mathrm{U}}, \end{aligned}$$

with $x_i^{\mathrm{L}} = [0.25, -0.05, -0.05, 0.001, 0.001, -0.2, -0.2, 0.015, 0.4, 0, 1, 0.04, 0.2]^{\mathrm{T}}$ and $x_i^{\mathrm{U}} = [0.31, 0.05, 0.05, 0.2, 0.2, 0.2, 0.2, 0.030, 4, \pi/2, 2, 0.5, 1.5]^{\mathrm{T}}$.

The minimum DBM values are set to 0.02 m and 0.05 m for SSP and DSP respectively. The decision variables reduce to twelve during DSP due to the absence of swing height parameters. The gait cycle consists of SSP and DSP. This study analyzed these phases separately to understand their distinct characteristics. The end state of SSP (position, velocity, acceleration) becomes the initial state for the subsequent DSP. This connection is essential for continuous motion. However, optimal parameters for each phase differ due to their distinct biomechanical constraints. During SSP, the swing leg requires specific velocity profiles to achieve desired swing height and forward progression. The support leg maintains stability with a smaller support polygon. These terminal velocities from SSP influence DSP's initial conditions. DSP operates under different constraints. Both feet contact the ground during DSP, creating a larger support polygon. This enables higher sagittal velocities and allows different lateral motion requirements for torso transfer between legs. Optimizing phases separately allows each to select parameters suited to its requirements. In practice, these phases can be joined later using optimal gait parameters from both phases. This maintains continuity at phase boundaries through shared velocity and position constraints.

5.1 EMO Algorithm Implementation

Two prominent EMO algorithms address the multi-objective optimization problem: the NSGA-II and MOPSO. These algorithms incorporate specialized mechanisms for Pareto optimality handling and solution diversity maintenance across the objective space. NSGA-II employs a population of 100 individuals evolved over 200 generations. The algorithm utilizes SBX with distribution index $\eta_c = 20$ and crossover probability $P_c = 0.9$. Polynomial mutation operates with distribution index $\eta_m = 50$ and mutation probability $P_m = 0.1$. MOPSO maintains identical population size and generations for fair comparison, with inertia weight $w = 0.5$, cognitive and social acceleration coefficients $c_1 = c_2 = 2.0$, and mutation probability $P_m = 0.1$.

5.2 Algorithm Performance Comparison

Figure 8a presents PFs obtained from both algorithms during SSP optimization. NSGA-II (blue) demonstrates uniform solution distribution, while MOPSO (red) identifies extreme solutions with superior stability characteristics absent in NSGA-II solutions. However, direct algorithm comparison proves challenging without reference PFs. To address this limitation, a reference set construction methodology combines non-dominated solutions from multiple independent runs. Six runs per algorithm generate twelve separate PFs. Solution aggregation followed by non-dominated sorting produces the approximated reference front (green), serving as a comparison baseline for performance evaluation.

The SSP reveals a distinct knee region (KR) where marginal improvements in one objective cause significant deterioration in the other (highlighted rectangular region).

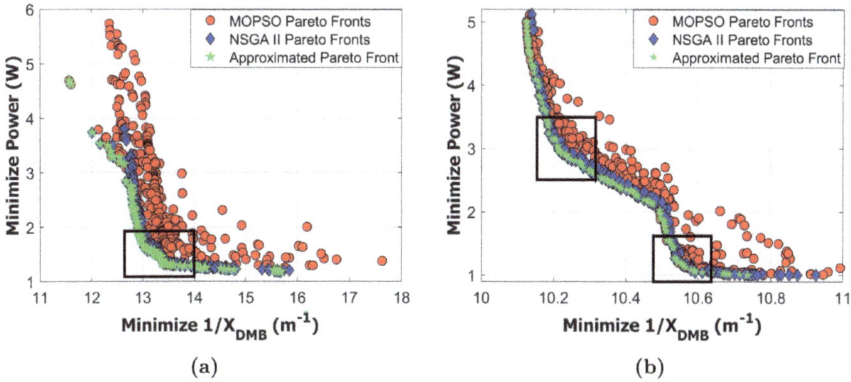

Fig. 8 PFs from NSGA-II (blue) and MOPSO (red) algorithms during **a** SSP and **b** DSP optimization. The rectangular regions highlight the knee areas offering superior trade-offs. SSP offers one KR while DSP reveals two distinct KRs offering superior trade-offs [5]

This KR represents solutions of particular interest for DMs, offering superior trade-offs compared to other PF regions. The DSP optimization exhibits increased complexity, as shown in Fig. 8b. Two distinct KRs emerge, indicating multiple favorable trade-off areas between objectives. This complexity results from the enlarged support polygon during DSP, where both feet contact the ground, providing enhanced stability options. The distinct PFs for SSP and DSP confirm these phases require different parameter selections. SSP exhibits a single KR reflecting the constrained nature of SSP. Power-stability trade-offs during SSP depend on the smaller support polygon and the need for precise swing leg control. In contrast, DSP reveals two distinct KRs. One region represents solutions where the robot walks at a faster pace, enhancing stability while consuming higher power. The second region corresponds to slower movement with lower energy requirements. The enlarged support polygon during DSP permits wider ranges of initial sagittal velocities compared to SSP. Final sagittal velocities during DSP are significantly lower, facilitating smooth transition to the next SSP phase with minimal jerk. DSP demonstrates different parameter sensitivity. Lateral velocities must be negative for torso transfer from positive to negative y-direction. SSP maintains near-zero lateral velocities for balance. These different front geometries and parameter ranges validate treating SSP and DSP as separate optimization problems. The reference set methodology offers advantages for algorithm comparison by eliminating *a priori* knowledge requirements of true PFs and providing unbiased comparison baselines derived from collective algorithm performance.

Dynamic Performance Analysis Using Running IGD+ Metrics

Traditional EMO algorithm comparisons focus exclusively on final generation results, overlooking dynamic convergence behavior during optimization. This limitation prevents understanding of algorithm stagnation, convergence patterns, and solution quality evolution. Dynamic performance evaluation using running metrics provides comprehensive insights into algorithm behavior throughout the optimization trajectory. The Inverted Generational Distance Plus (IGD+) indicator effectively tracks algorithm progress generation-by-generation, measuring both convergence and diversity relative to the reference set. Lower IGD+ values indicate superior performance, representing closer reference front approximation with improved solution distribution.

Figure 9a illustrates running IGD+ evolution for both algorithms during SSP optimization. NSGA-II demonstrates consistent improvement with steady IGD+ reduction and occasional minor fluctuations, suggesting effective exploration-exploitation balance without premature stagnation. MOPSO exhibits rapid initial improvement followed by stagnation between generations 20–70, indicating potential premature convergence or diversity loss. DSP dynamics differ significantly, as shown in Fig. 9b. MOPSO achieves rapid early convergence within 30 generations but subsequently stagnates with minimal late-generation progress. NSGA-II maintains

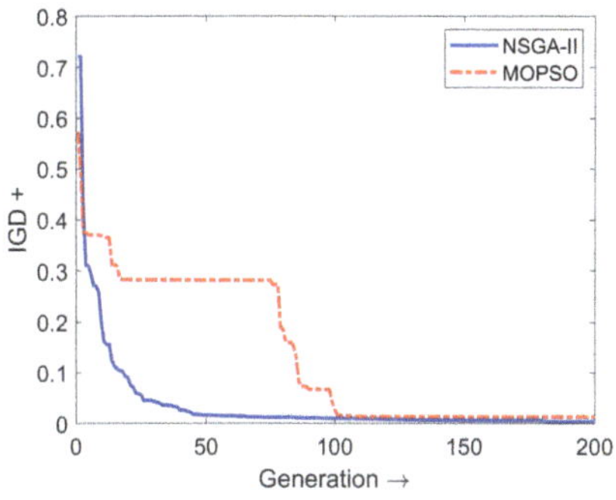

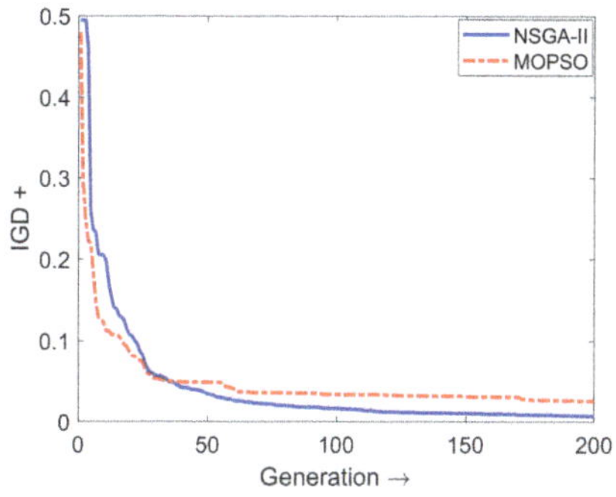

Fig. 9 Running IGD+ evolution for NSGA-II and MOPSO during **a** SSP and **b** DSP optimization. The plots reveal convergence patterns and stagnation periods, with NSGA-II demonstrating superior long-term performance, particularly during DSP optimization [5]

consistent improvement throughout optimization, achieving superior final performance despite slower initial progress. This analysis reveals NSGA-II's effectiveness for complex constraint structures and multiple KRs characteristic of DSP optimization. An extended evaluation by Gupta et al. [28] compared four algorithms namely NSGA-II, Competitive Mechanism-based MOPSO (CMOPSO), Dynamic Constrained NSGA-III (DCNSGA-III), and Constrained Multi-Objective Evolutionary Algorithm based on Decomposition (CMOEA/D) using running IGD and hypervolume metrics for NAO robot gait optimization, demonstrating NSGA-II's superior reliability and consistency. Their analysis revealed distinct behavioral patterns during optimization, with CMOPSO showing competitive initial performance but occasional stagnation, DCNSGA-III exhibiting constraint-handling difficulties, and CMOEA/D demonstrating poor convergence characteristics requiring significantly more computational resources.

5.3 Knee-Based Multi-objective Optimization

EMO algorithms generate a large number of solutions on PFs representing trade-offs between conflicting objectives. This sheer number can overwhelm DMs. An emerging paradigm focuses on identifying specific KR that represent superior trade-offs between objectives. Outside these regions, a small improvement in one objective results in significant deterioration in others, making KRs attractive for practical decision-making, as illustrated in Fig. 10. DMs typically prefer solutions providing balanced performance across multiple objectives without excessive compromise in any criterion.

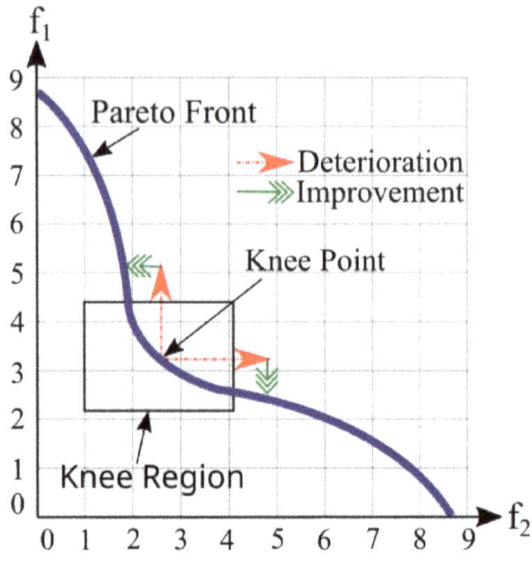

Fig. 10 Knee region concept in bi-objective optimization showing the PF with marked knee point where significant curvature changes occur, representing superior trade-offs between objectives f_1 and f_2

Angle-Based Knee Detection

The angle-based approach focuses on KRs by analyzing geometric properties of the PF by examining the slopes of lines connecting individual solutions with their neighbors, without requiring prior knowledge of DM preferences, as shown in Fig. 11a. For a solution $\mathbf{x}_i$ positioned between neighbors $\mathbf{x}_{i-1}$ and $\mathbf{x}_{i+1}$ on a sorted PF, the angle α represents the directional change in objective space. Solutions with angles significantly exceeding 180° indicate potential knee points where small improvements in one objective cause substantial deterioration in another. The method can be enhanced by considering multiple neighbors simultaneously, calculating extended angles β, γ, and δ by examining two, three, and four neighbors respectively. The four-angle measurement approach considers $\max(\alpha, \beta, \gamma, \delta)$, providing more aggressive KR detection by reducing sensitivity to local irregularities.

Consider a bi-objective problem $f_1 = x$ with $x \in [0, 1]$ and $f_2 = (1-x)^2 + 0.1\sin(8\pi x)$ having 50 uniformly distributed solutions exhibiting natural curvature. Figure 11b shows that the angle-based method successfully identifies KRs where PF curvature is most pronounced. The method calculates angles for interior points, with larger angles (marked by larger circles) indicating stronger knee characteristics. This geometric approach provides intuitive knee identification based on PF shape alone, making it particularly suitable for problems where DM preferences are unavailable or difficult to articulate.

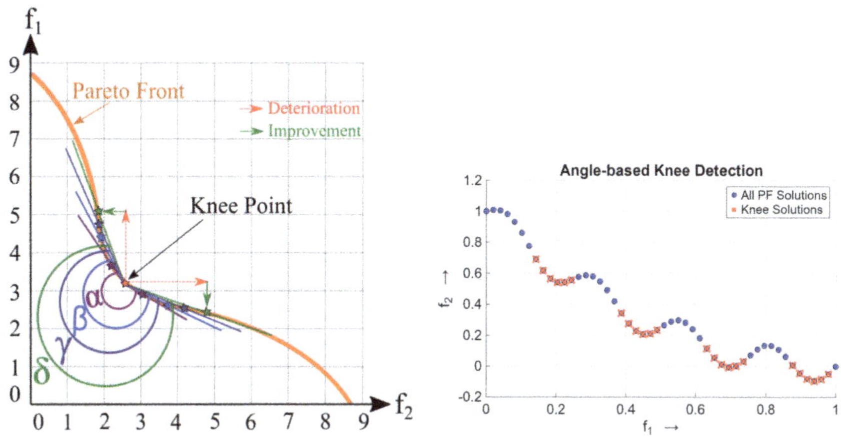

(a) (b)

Fig. 11 Angle-based knee detection results on DO2DK benchmark problem. **a** Calculation of angles α, β, γ, and δ by taking 1, 2, 3 and 4 neighbors, respectively. **b** Angle-based knee detection on simple benchmark problem. Solutions with angles exceeding 180.2° are identified as knee points (larger red markers), demonstrating the method's ability to capture regions of high PF curvature

Implementation on DO2DK Benchmark Problem

Angle-based approach is validated using the DO2DK benchmark problem, mathematically formulated in Eqs. (13) and (14). Parameter K controls the number of knee points, n represents decision variables, and s determines PF skewness. For this study, parameters $K = 5, n = 30$, and $s = 1$ are set to generate five distinct KRs. This tests the algorithm's ability to identify and focus on multiple knee areas simultaneously. To focus on KRs, NSGA-II's crowding distance mechanism is replaced with the angle-based method. In each generation, non-dominated solutions are sorted by the first objective, and angles are calculated for interior points. Solutions with larger angles receive higher selection priority, directing the search toward KRs. This modification transforms NSGA-II from a diversity-maintaining algorithm to a knee-focused optimizer while preserving its elitist structure and convergence properties. NSGA-II is configured with population size of 200 evolved over 200 generations, crossover distribution index $\eta_c = 20$ with probability $P_c = 0.9$, and mutation distribution index $\eta_m = 50$ with probability $P_m = 0.1$.

$$\min f_1(\mathbf{x}) = g(\mathbf{x})r(\mathbf{x}_1)\left(\sin\left(\frac{\pi \mathbf{x}_1}{2^{s+1}} + \left(1 + \frac{2^s - 1}{2^{s+2}}\right)\pi\right) + 1\right), \tag{13}$$

$$\min f_2(\mathbf{x}) = g(\mathbf{x})r(\mathbf{x}_1)\left(\cos\left(\frac{\pi \mathbf{x}_1}{2} + \pi\right) + 1\right), \tag{14}$$

where

$$g(\mathbf{x}) = 1 + \frac{9}{n-1}\sum_{i=2}^{n}\mathbf{x}_i,$$

$$r(\mathbf{x}_1) = 5 + 10(\mathbf{x}_1 - 0.5)^2 + \frac{1}{K}\cos(2K\pi\mathbf{x}_1)\cdot 2^{\frac{s}{2}},$$

subject to

$$0 \le \mathbf{x}_i \le 1, \quad i = 1, 2, \ldots, n.$$

Figure 12 illustrates the progressive behavior of both methods across 200 generations. Figure 12a shows Generation 1, where both methods start with random populations without selection pressure for efficient regions or diversity maintenance. Figure 12b shows Generation 25, where both methods produce solutions with improved convergence due to rank-based selection. The knee-based method shows more concentrated clustering compared to crowding distance, with many solutions gathering around KRs. The KRs represent the most efficient trade-off areas. Some KRs show heavy concentration while others require more generations. The crowding distance maintains diversity across the entire PF. Figure 12c shows Generation 50, where continued convergence improvement occurs in both methods. Both methods have approached the near-true PF for DO2DK. The crowding distance maintains good diversity, whereas the angle-based method focuses on KRs with less uniform coverage. The knee-based method achieved better convergence than crowding distance at Generation 25, with solutions visible in high-curvature regions outside the five primary knees. By Generation 50, both methods reach similar convergence levels. Some KRs exhibit clear separation while others still cover larger areas, remaining less diverse than the crowding distance method. Figure 12d shows Generation 75, where the angle-based method shows very few solutions at extremes. The lower right region of PF still covers a larger region, likely because the five knees provide substantial local information that limits global information about PF during angle-based selection. Figure 12e–g show Generations 100, 125, and 150 respectively, revealing five distinct knee zones with progressive clarity. The separation between KRs emerges with each generation. The crowding distance method has already found the most diverse regions coupled with good convergence. The rightmost KR in the lower-right still covers considerable area. Figure 12h shows Generation 200, where complete identification of all five KRs occurs. The solutions cluster tightly around maximum curvature points, effectively avoiding flat PF regions. The standard NSGA-II with crowding distance maintains diversity across the entire PF.

This progressive evolution demonstrates how the angle-based method automatically identifies and concentrates on KRs without *a priori* preference information. By calculating angles at each solution and preferring larger angles during selection, the algorithm progressively abandons low-curvature regions where small objective improvements require large sacrifices in the other objective. The five distinct knee zones emerge naturally through the geometric properties of the PF, with solutions

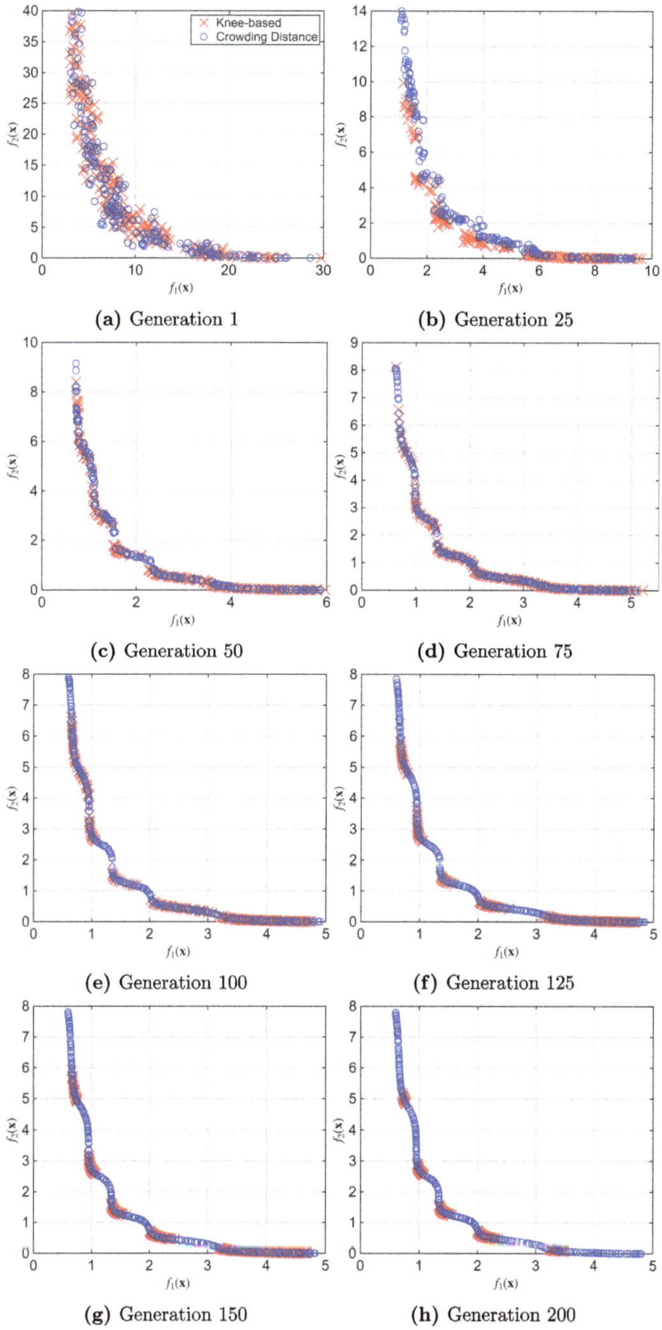

Fig. 12 Progressive evolution of NSGA-II with angle-based knee detection on DO2DK benchmark problem. Blue circles show standard NSGA-II with crowding distance maintaining diversity across the entire PF, while red crosses show knee-based NSGA-II progressively focusing on five knee regions

concentrating where trade-offs are most favorable. This reduces decision-maker burden from evaluating hundreds of uniformly distributed solutions to selecting from approximately 30–40 solutions in the most efficient trade-off zones.

The DO2DK benchmark problem provides controllable parameters for the number of knees (K) and skewness (s), offering flexibility for testing algorithms under varying complexities. While other benchmark suites such as Zitzler–Deb–Thiele (ZDT), Deb–Thiele–Laumanns–Zitzler (DTLZ), or Walking Fish Group (WFG) could provide broader validation across diverse PF geometries, DO2DK's configurable characteristics allow focused analysis of the algorithm's sensitivity and accuracy under varying conditions. The subsequent application to humanoid robot gait generation (Sect. 5.3) demonstrates effectiveness on a real-world optimization problem.

Implementation on Humanoid Robot Gait Generation

NSGA-II algorithm is employed for KR detection during both SSP and DSP phases. Similar to the DO2DK problem, the angle-based knee detection methodology is applied to the humanoid robot gait optimization problem for both SSP and DSP. NSGA-II parameters remain consistent with Sect. 5: population size of 152, 200 generations, crossover distribution index $\eta_c = 20$ with probability 0.9, and mutation distribution index $\eta_m = 50$ with probability 0.1. The crowding distance operator is replaced with the angle-based metric to focus the search on KRs. Non-dominated solutions in each generation are preferred based on higher angle values. Figure 13 presents the results for both gait phases.

For SSP, Fig. 13a shows the complete PF obtained using standard NSGA-II with crowding distance, revealing solutions distributed across the entire trade-off surface. Figure 13b shows knee-based NSGA-II results, where solutions concentrate in a single distinct KR. This region represents critical trade-offs between power consumption and stability: moving away from this knee in either direction causes one objective to deteriorate significantly faster than the other improves. The single KR during SSP reflects the constrained nature of single-leg support, where the limited support polygon restricts the range of viable gait parameters. Unlike DO2DK, solutions here are not heavily concentrated in the KR. The algorithm finds fewer solutions in other regions because this real-world problem exhibits non-uniform curvature. The angle-based method sometimes identifies larger angles at scattered locations across the PF, retaining solutions that provide better local trade-offs compared to their neighborhoods, rather than concentrating exclusively in KRs as observed in DO2DK. A real-world problem has nonuniform curvature unlike DO2DK.

For DSP, Fig. 13c shows the complete PF exhibiting greater complexity than SSP due to the enlarged support polygon when both feet contact the ground. Figure 13d reveals that knee-based NSGA-II identifies two distinct KRs during DSP. The first KR corresponds to dynamic stability solutions requiring higher power consumption to achieve rapid weight transfer between legs. The second KR represents static stability solutions with lower energy requirements, exploiting the inherently stable dual-support configuration to enable slower movements while maintaining adequate

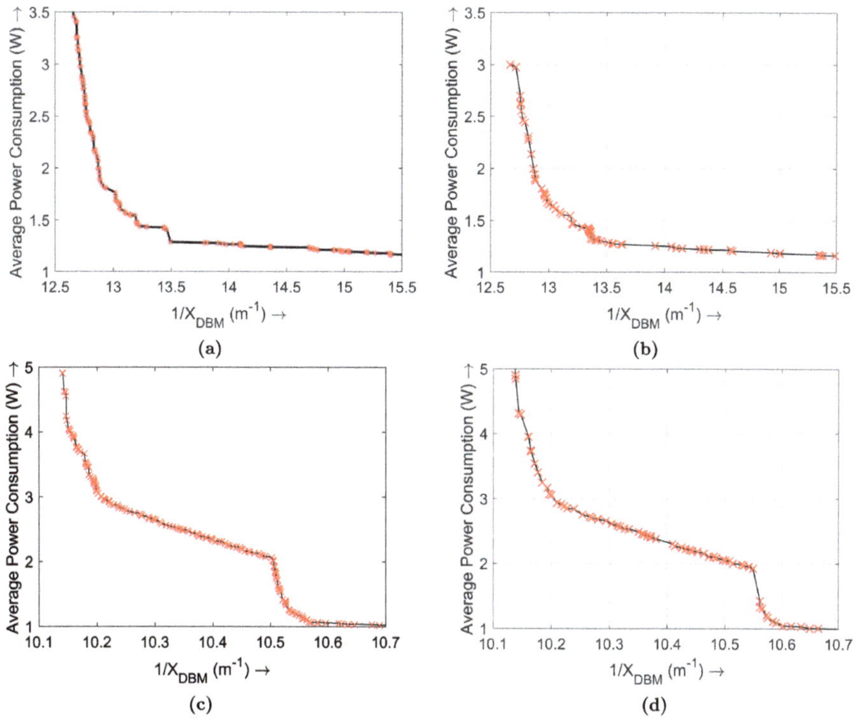

Fig. 13 Knee-based optimization results for humanoid robot gait generation. Single support phase: **a** complete PF using crowding distance, **b** knee region solutions using angle-based detection. Double support phase: **c** complete PF using crowding distance, **d** two knee regions identified using angle-based detection [6]

stability margins. The presence of two KRs validates the optimization approach: rather than providing hundreds of solutions spanning the entire trade-off space, the angle-based method identifies more solutions in KRs offering the most favorable trade-offs for practical gait implementation.

These results demonstrate that knee-based optimization effectively reduces the decision-making burden by focusing on PF regions most relevant to practical robotics applications. The identified knee solutions provide balanced performance across both objectives, avoiding extreme solutions that sacrifice one objective excessively. This focused approach proves particularly valuable for humanoid robot gait generation, where implementable gaits must satisfy multiple competing requirements within strict physical constraints. Previous work in [6] also demonstrated the utility-based approach to detect knee during SSP and DSP scenarios. The two-objective formulation presented in this chapter can be extended to many-objective optimization involving three or more objectives simultaneously. Gupta et al. [29] optimized four conflicting objectives, namely power consumption, stability, torque fluctuation, and gait cycle time, using NSGA-III with Type I and Type II robustness approaches for the NAO robot. Their work demonstrated the effectiveness of many-objective evolutionary algorithms in handling complex bipedal locomotion problems.

6 Conclusions and Future Directions

This chapter has presented a comprehensive exploration of emerging techniques in EC, demonstrating their effectiveness through application to the challenging problem of humanoid robot gait generation. The integration of novel comparison methodologies, advanced performance indicators, and knee-based optimization approaches have shown significant advancement in the field of EC. The comprehensive statistical comparison methodology for single-objective optimization algorithms has addressed fundamental challenges of fair performance evaluation in stochastic optimization environments. This approach, validated through eleven independent runs, has revealed that PSO consistently outperforms GA in achieving superior results across both identical and different initial population sets for single-objective version of the problem. The narrow performance bands exhibited by PSO across multiple runs demonstrate superior algorithm reliability compared to GA. For the multi-objective version of the problem, NSGA-II has demonstrated better performance when compared with MOPSO. MOPSO has exhibited stagnation periods revealed through running performance metrics. This insight has profound implications for algorithm selection based on available computational resources and desired optimization outcomes. Moreover, the dynamic evaluation using running IGD+ metrics has revealed intermediate algorithm behaviors (particularly stagnation phases) that remain invisible when comparing only the final-generation PFs. Finally, the knee-based optimization methodology has successfully reduced decision-maker burden by focusing on specific PF regions offering superior trade-offs. The comparative analysis of angle-based and utility-based knee detection methods has revealed that while angle-based approaches offer computational efficiency, utility-based methods provide superior solution distribution and comprehensive KR coverage. The application to humanoid robot gait optimization has demonstrated practical effectiveness, with the SSP exhibiting one distinct KR while the DSP revealing two, reflecting the underlying behavior of bipedal locomotion.

This chapter presents methodological advances in evolutionary optimization techniques demonstrated through the humanoid gait generation problem. The optimization framework is validated through comprehensive kinematic and dynamic simulations based on NAO robot specifications. The primary contribution lies in the comparison methodologies, running performance metrics, and knee-based optimization techniques that are broadly applicable across optimization domains. The simulated environment provides controlled conditions essential for systematic algorithm comparison and performance analysis. Physical hardware implementation remains a valuable direction for future work to validate the optimized gaits under real-world conditions.

The integration of emerging techniques such as running performance metrics, reference PF construction for real-world problems, and angle-based knee optimization represents a significant advancement in how EC problems are approached and solved. The extension of the presented methodologies to many-objective optimization problems involving four or more objectives presents different challenges and opportunities. While running performance metrics offer insights into algorithm stagnation, future work should explore the possibility of real-time algorithm termination when improvement ceases. Machine learning techniques, particularly reinforcement learning approaches, could be employed to guide the search process by learning from optimization trajectories, thereby avoiding stagnation regions identified through running performance metrics. The angle-based knee detection method demonstrated in this chapter focuses on bi-objective problems. Extension to many-objective optimization requires alternative approaches for identifying knee regions in higher-dimensional objective spaces. Furthermore, the development of adaptive termination criteria based on running metrics could reduce computational requirements while ensuring solution quality, making evolutionary algorithms more practical for resource-constrained environments.

References

1. Zhang, Q., Hu, Z., Hong, N., Su, Q.: A fixed point evolution algorithm based on expanded Aitken rapid iteration method for global numeric optimization. Math. Comput. Simul. **229**, 288–303 (2025)
2. Gupta, P., Kumar, V., Pratihar, D.K., Deb, K.: Multi-objective optimization of rotational magnetorheological abrasive flow finishing process. In: Nanofinishing of Materials for Advanced Industrial Applications, pp. 13–33. CRC Press (2023)
3. ur Rehman, J., Ulum, M.S., Shaffar, A.W., Hakim, A.A., Abdullah, Z., Al-Hraishawi, H., Chatzinotas, S., Shin, H., et al.: Evolutionary algorithms and quantum computing: recent advances, opportunities, and challenges. IEEE Access (2025)
4. Gupta, P., Pratihar, D.K., Deb, K.: Analysis and optimization of gait cycle of 25-DoF NAO robot using particle swarm optimization and genetic algorithms. Int. J. Humanoid Robot. **21**(02), 2350011(1)–2350011(44) (2024)
5. Gupta, P., Pratihar, D.K., Deb, K.: A comparative study of pareto front of optimal solution set for NAO robot's gait optimization using the dominance move indicator based on mixed integer programming. In: Sharma, D., Gautam, S.S., Mankodi, T.K., Saha, U.K. (eds.) Advances in Multidisciplinary Design, Analysis and Optimization, pp. 427–435. Springer Nature Singapore, Singapore (2025)
6. Gupta, P., Pratihar, D.K., Deb, K.: A knee-based multi-objective optimization for gait cycle of 25-DoF NAO humanoid robot. In: Pant, M., Deep, K., Nagar, A. (eds.) Proceedings of the 12th International Conference on Soft Computing for Problem Solving, pp. 47–62. Springer Nature Singapore, Singapore (2024)
7. Song, Y., Wu, Y., Guo, Y., Yan, R., Suganthan, P.N., Zhang, Y., Pedrycz, W., Das, S., Mallipeddi, R., Ajani, O.S., et al.: Reinforcement learning-assisted evolutionary algorithm: a survey and research opportunities. Swarm Evol. Comput. **86**, 101517 (2024)
8. Zhang, Y., Cheng, L., Chen, G., Alghazzawi, D.: Evolutionary computation in bioinformatics: a survey. Neurocomputing **591**, 127758 (2024)

9. Kashyap, A.K., Parhi, D.R.: Implementation of intelligent navigational techniques for inter-collision avoidance of multiple humanoid robots in complex environment. Appl. Soft Comput. **124**, 109001 (2022)
10. Kashyap, A.K., Parhi, D.R., Pandey, A.: Multi-objective optimization technique for trajectory planning of multi-humanoid robots in cluttered terrain. ISA Trans. **125**, 591–613 (2022)
11. Muni, M.K., Parhi, D.R., Kumar, P.B.: Improved motion planning of humanoid robots using bacterial foraging optimization. Robotica **39**, 123–136 (2021)
12. Abedzadeh Maafi, R., Etemadi Haghighi, S., Mahmoodabadi, M.J.: Pareto optimal design of a fuzzy adaptive sliding mode controller for a three-link model of a biped robot via the multi-objective improved team game algorithm. J. Braz. Soc. Mech. Sci. Eng. **44**(9), 428 (2022)
13. Raj, M., Semwal, V.B., Nandi, G.C.: Multiobjective optimized bipedal locomotion. Int. J. Mach. Learn. Cybern. **10**, 1997–2013 (2019)
14. Wolpert, D., Macready, W.: No free lunch theorems for optimization. IEEE Trans. Evol. Comput. **1**(1), 67–82 (1997)
15. Mitchell, M.: An Introduction to Genetic Algorithms. MIT Press (1998)
16. Kennedy, J., Eberhart, R.: Particle swarm optimization. In: Proceedings of ICNN'95—International Conference on Neural Networks, vol. 4, pp. 1942–1948 (1995)
17. Van Veldhuizen, D.A., Lamont, G.B., et al.: Evolutionary computation and convergence to a pareto front. In: Late Breaking Papers at the Genetic Programming 1998 Conference, pp. 221–228. Citeseer (1998)
18. Ishibuchi, H., Masuda, H., Tanigaki, Y., Nojima, Y.: Modified distance calculation in generational distance and inverted generational distance. In: Evolutionary Multi-Criterion Optimization: 8th International Conference, EMO 2015, Guimarães, Portugal, 29 Mar–1 Apr 2015. Proceedings, Part II 8, pp. 110–125. Springer (2015)
19. Das, I.: On characterizing the "knee" of the pareto curve based on normal-boundary intersection. Struct. Optim. **18**(2), 107–115 (1999)
20. Blank, J., Deb, K.: A running performance metric and termination criterion for evaluating evolutionary multi- and many-objective optimization algorithms. In: 2020 IEEE Congress on Evolutionary Computation (CEC), pp. 1–8. IEEE (2020)
21. Yu, G., Ma, L., Jin, Y., Du, W., Liu, Q., Zhang, H.: A survey on knee-oriented multiobjective evolutionary optimization. IEEE Trans. Evol. Comput. **26**(6), 1452–1472 (2022)
22. Fragakis, N., Trichopoulos, G., Caridakis, G.: Empowering education with intelligent systems: exploring large language models and the NAO robot for information retrieval. Electronics **14**(6), 1210 (2025)
23. Dan, A., Saha, S., Krishna, K.R.: A review on the stability of biped robots. J. Indian Inst. Sci. 1–31 (2025)
24. Vukobratović, M., Borovac, B.: Zero-moment point—thirty five years of its life. Int. J. Humanoid Rob. **1**, 157–173 (2004)
25. NAO Version 6 Description. https://www.aldebaran.com/en/nao. Accessed 22 June 2023
26. Deb, K., Pratap, A., Agarwal, S., Meyarivan, T.: A fast and elitist multiobjective genetic algorithm: NSGA-II. IEEE Trans. Evol. Comput. **6**(2), 182–197 (2002)
27. Coello, C., Pulido, G., Lechuga, M.: Handling multiple objectives with particle swarm optimization. IEEE Trans. Evol. Comput. **8**(3), 256–279 (2004)
28. Gupta, P., Pratihar, D.K., Deb, K.: Dynamic performance evaluation of evolutionary multi-objective optimization algorithms for gait cycle optimization of a 25-DoFs NAO humanoid robot. Swarm Evol. Comput. **99**, 102144 (2025)
29. Gupta, P., Pratihar, D.K., Deb, K.: Many-objective robust gait optimization for a 25-DoFs NAO robot using NSGA-III. Eng. Optim. 1–39 (2025)

Optimization Models in Nurse Scheduling—A Review of Last Five Years

Rutvij Tole and Millie Pant

Abstract In a bustling hospital ward, the heart of healthcare beats in rhythms defined by nurses along with their shifts, their rest, their presence. But behind the calm of patient care lies a complex puzzle: how to ensure the right number of skilled nurses is available at the right time, every time. This paper takes readers through the evolving landscape of nurse scheduling, exploring 37 research studies in the Web of Science database published between 2020 and 2024. From hospitals overwhelmed by COVID-19 to digital systems learning to adapt in real-time, these studies reveal a silent yet critical battle, optimizing care through smarter rosters. This review categorizes the strategies from traditional optimization models to cutting-edge AI-driven systems and spotlights key challenges such as cost, real-time adaptability, and workforce satisfaction. The goal is simple yet profound: to support sustainable staffing that puts both patients and nurses at the center.

Keywords Nurse scheduling · Staff scheduling · NSP · Nurse rostering

1 Introduction

In the corridors of any hospital, nurses are the silent engines keeping the healthcare system running by administering care, offering comfort, and responding swiftly to medical emergencies. But behind this seamless coordination lies a complicated puzzle: how do hospitals ensure that the right nurse is available at the right place, at the right time on every day, every shift? This challenge is known as nurse scheduling.

R. Tole · M. Pant
Department of Applied Mathematics and Scientific Computing, Indian Institute of Technology Roorkee, Roorkee, Uttarakhand, India
e-mail: rutvij_t@amsc.iitr.ac.in

M. Pant (✉)
Mehta Family School for Data Science and Artificial Intelligence, Indian Institute of Technology Roorkee, Roorkee, Uttarakhand, India
e-mail: pant.milli@as.iitr.ac.in

S. Mukhopadhyay et al. (eds.), *Contemporary Advancement in Evolutionary Multi-objective Optimization*, Genetic and Evolutionary Computation,
https://doi.org/10.1007/978-981-95-7908-2_5

In the context of a hospital management system (HMS), nurse scheduling refers to the process of creating and managing work schedules for nurses to ensure that patient care needs are met efficiently while considering various operational, legal, and human factors [1]. It is a critical aspect of hospital management because it directly impacts patient care quality, nurse well-being, and hospital efficiency. A single mistake in scheduling can mean overworked staff, missed care, or unsafe conditions for both nurses and patients.

Nurse scheduling involves several key aspects (Fig. 1) that are vital to ensuring the delivery of high-quality patient care while maintaining operational efficiency and nurse satisfaction. One important focus is matching staffing levels with patient demand to ensure appropriate patient-to-nurse ratios, which are often regulated by laws or hospital policies. Effective scheduling must account for fluctuations in patient admissions, discharges, and varying care intensity across different units. Compliance with labor laws and regulations, such as limits on working hours, mandatory breaks, and legal patient-to-nurse ratios, is another critical element. At the same time, scheduling must incorporate nurse preferences for shifts and days off to improve job satisfaction and minimize burnout [2]. Balancing shift coverage across units is essential to avoid under- or over-staffing while ensuring continuity of care by placing nurses with specialized skills in appropriate roles. Additionally, cost optimization plays a significant role, as scheduling should minimize unnecessary expenses, such as overtime or reliance on temporary staff, while still meeting patient care needs. Flexibility is increasingly important, with systems allowing for real-time adjustments to accommodate sudden changes in patient demand or emergencies, often using float nurses or cross-trained staff. These aspects collectively contribute to better patient outcomes, enhanced nurse well-being, and improved hospital operational efficiency.

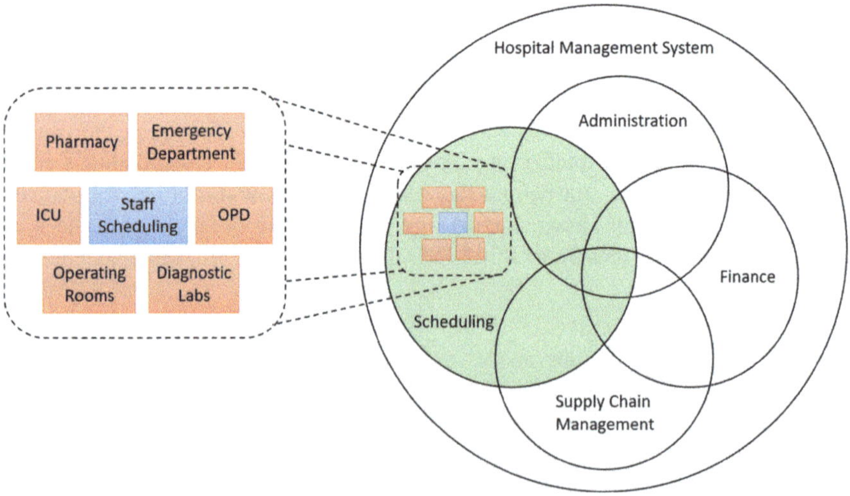

Fig. 1 Staff scheduling in context of hospital management system

Traditionally, nurse scheduling was done manually, a task often handled by head nurses or administrators using paper rosters or spreadsheets. But as hospitals have grown in size, complexity, and unpredictability, manual scheduling has proven inefficient and error-prone. Modern hospital management systems now incorporate software solutions [3] that automate scheduling using advanced algorithms. These systems are designed to consider multiple constraints, such as nurse preferences, regulatory requirements, and patient demand, allowing for more efficient and accurate scheduling. Optimization models, including mixed-integer programming and heuristic approaches, are frequently employed to generate schedules that balance cost, efficiency, and nurse satisfaction. Additionally, real-time adjustments are made possible by integrating patient admission data and staffing levels into scheduling systems, enabling dynamic updates to address unexpected changes in demand or emergencies. Artificial intelligence and machine learning are also increasingly utilized to predict staffing needs based on historical data and trends, further enhancing the adaptability and scalability of nurse scheduling processes. These tools and methods collectively support a more responsive and effective scheduling system [4], ensuring high-quality patient care and optimal resource utilization.

Nurse scheduling encompasses several interconnected aspects that are critical for the efficient functioning of a hospital. At its core, it involves aligning staffing levels with patient demand, ensuring that sufficient nurses are available to maintain optimal patient-to-nurse ratios, which directly affect care quality and safety. This process must account for variability in patient admissions, discharges, and the intensity of care required in different units. Additionally, nurse scheduling must comply with labor regulations, such as limits on working hours, mandatory rest periods, and legal standards for staffing ratios, to protect both patients and nurses. Considering nurse preferences is another vital aspect, as accommodating requests for specific shifts or days off improves job satisfaction, reduces burnout, and supports retention [5]. Effective scheduling also requires balancing shift coverage to prevent under- or over-staffing and ensuring continuity of care by assigning nurses with specialized skills to the appropriate units. Cost optimization is equally important, as scheduling decisions need to minimize unnecessary expenditures, such as overtime or reliance on temporary staff, while maintaining quality care. Moreover, flexibility is crucial in nurse scheduling, allowing for real-time adjustments to address unexpected demand fluctuations or emergencies, often using float nurses or cross-trained staff. These aspects collectively contribute to creating a well-functioning, sustainable, and patient-centered healthcare system.

Nurse scheduling is essential because it ensures the right number of skilled nurses are available to provide high-quality patient care while maintaining operational efficiency. Proper scheduling directly impacts patient outcomes by meeting care demands and maintaining optimal patient-to-nurse ratios. It also plays a critical role in supporting nurse satisfaction by considering preferences and reducing burnout, which helps retain a skilled workforce. Additionally, effective scheduling optimizes hospital resources, minimizes unnecessary costs like overtime, and ensures compliance with labor regulations. In dynamic healthcare environments, flexible scheduling is crucial for responding to emergencies and fluctuations in patient demand, making it a cornerstone of efficient and sustainable hospital management. Based on the abovementioned aspects, the present study tries to answer the following research questions:

1. What are the key methods and approaches employed in nurse scheduling research from 2020 to 2024, and how do they compare in terms of effectiveness and practicality for nurse scheduling?
2. How effectively do current nurse scheduling models incorporate real-time data and dynamic changes in patient demand?
3. What emerging trends, such as the use of AI or workforce flexibility, are influencing nurse scheduling research, and what gaps remain for future exploration?

Compared to earlier review articles by Ngoo et al. [6], Little et al. [7], and Zhang et al. [8], this paper introduces several important updates and extensions. While previous works provided valuable insights at the time of their publication, this paper offers a generalized model description, a structured and detailed literature synthesis, an expanded tabular comparison of problem characteristics and methods, an in-depth bibliometric and country-wise research analysis, and new perspectives shaped by the latest developments in the field. Notably, a dedicated section on Reinforcement Learning approaches is included, addressing advancements that have gained prominence only in recent years and were not available during the preparation of earlier reviews. These enhancements position this paper as a timely and comprehensive resource that captures both foundational knowledge and emerging trends. Table 1 gives the comparative information about different aspects in recent review articles.

Remember, effective nurse scheduling is no longer just about efficiency; it's about resilience and adaptability. Hospitals must plan for surging patient demand, sudden staff shortages, or emergency events, all while upholding standards of quality care. Moreover, with increasing awareness around staff well-being and mental health, incorporating nurse preferences has become a priority, not an afterthought.

The content of this paper has been organized in the following systematic way. The basic concept on nurse scheduling and a generalized model along with is discussed in Sect. 2. Research approach, literature search, and related work of this review paper

Table 1 Novelty chart for the review article

Paper characteristics	This article	Ngoo et al. [6]	Little et al. [7]	Zhang et al. [8]
Generalized model description	✔	✔	✘	✘
Detailed literature review	✔	✔	✔	✘
Tabular comparison of problem characteristics and tools used	✔	✔	✘	✔
Detailed bibliometric analysis	✔	✘	✘	✘
Country-wise research analysis	✔	✔	✘	✘
A dedicated section for Reinforcement Learning	✔	✘	✘	✘
Country specific literature	✘	✘	✔	✘

have been discussed in the third section. Section 4 discusses bibliometric analysis. Section 5 is a special section which involves a discussion regarding Reinforcement Learning for nurse scheduling. The discussion related to present article is given in Sect. 6. Section 7 provides concluding remarks, limitations, and future scope.

2 Nurse Scheduling Model

When a senior nurse in the ICU glanced at her weekly schedule, she felt a familiar mix of fatigue and resignation which was an another stretch of back-to-back night shifts. Despite her experience and seniority, the system didn't seem to recognize her preferences or past workload. Behind that frustrating schedule lay a much deeper issue: the absence of a model smart enough to balance patient care needs with human limits. This is exactly what nurse scheduling models try to solve.

Nurse scheduling models are frameworks and methodologies developed to address the complexities of allocating shifts to nurses while balancing various constraints and objectives. These models typically include optimization techniques like linear programming, mixed-integer programming, and goal programming, which focus on achieving cost efficiency and meeting staffing requirements. Heuristic models such as local search, Greedy Algorithms, and Variable Neighborhood Search, and metaheuristic models, such as genetic algorithms, particle swarm optimization, and simulated annealing, are widely used to find near-optimal solutions for large-scale scheduling problems. Dynamic models incorporate real-time data to adapt schedules as conditions change, whereas workforce flexibility models leverage float nurses and cross-training to enhance adaptability. Multi-objective models aim to balance competing goals such as cost minimization, nurse satisfaction, and quality of care, often making trade-offs to meet these objectives. These diverse models play a critical role in ensuring efficient hospital operations, improving nurse well-being, and delivering high-quality patient care. Diversification of these models is represented in Fig. 2.

At first glance, a nurse schedule might look like a simple grid of names and shifts. But beneath that surface lies a highly structured mathematical system a generalized nurse scheduling model designed to make sense of countless real-world constraints and decisions. This model is built on four foundational elements: variables, parameters, objective functions, and constraints. Variables determine which nurse is assigned to which shift. Parameters describe fixed inputs like staffing needs, shift durations, or nurse availability. Objective functions define what the schedule aims to achieve whether that's minimizing cost, reducing overtime, or maximizing nurse satisfaction. And constraints ensure the schedule is both legal and humane: Nurses can't work beyond a certain number of hours, rest days must be honored, and coverage must be maintained. Mathematical model is given below, and the diagrammatic structure of a general nurse scheduling model is given in Fig. 3.

Across the studies reviewed in this paper, these components appear in various combinations, but the core idea remains the same: create a schedule that balances

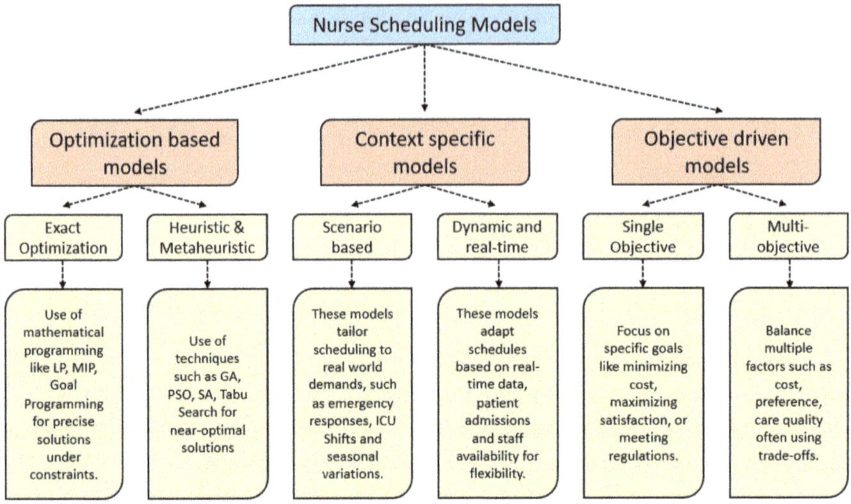

Fig. 2 Classification of nurse scheduling models

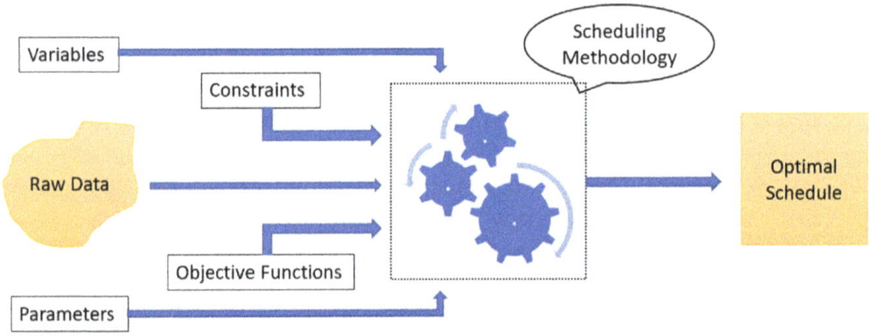

Fig. 3 General model for nurse scheduling

operational needs with human realities. In recent years, there's been a clear shift toward multi-objective formulations, where different goals like efficiency, fairness, and flexibility are woven into a single composite objective function. This reflects the growing understanding that nurse scheduling isn't a one-size-fits-all problem it's a delicate trade-off, and the best solutions are those that adapt to context while protecting both patient care and staff well-being.

Objective Functions

Constraints

Staffing level requirement ensures that each shift has the required number of nurses. Nurse assignment constraint ensures that a nurse can only be assigned to a shift if

Table 2 Objective functions

Objective	Mathematical form
Minimize total cost	min. $Z_1 = \sum_i \sum_j C_i \cdot x_{ij}$
Minimize total working hours	min. $Z_2 = \sum_i y_i$
Maximize nurse preference satisfaction	max. $Z_3 = \sum_i \sum_j P_i . x_{ij}$

Table 3 Constraints

Constraint	Mathematical form
Staffing level constraint	$\sum_i x_{ij} = S_j, \ \forall j$
Nurse assignment constraint	$x_{ij} \leq A_i, \ \forall i, \ \forall j$
Maximum working hours for nurses	$y_i = \sum_j T_j . x_{ij}, \ \forall i$
Maximum hours constraint	$y_i \leq W_i, \ \forall i$
Binary decision variable	$x_{ij} \in \{0, 1\}, \ \forall i, \ \forall j$

they are available. Maximum hours constraint ensures that no nurse works more than their maximum allowable hours.

Variables

Parameters

So, the overall mathematical problem can be compactly written as follows:

Let x_{ij} be a binary decision variable such that,

Table 4 Variables

Variable	Description
x_{ij}	Binary decision variable, 1 if nurse i is assigned to shift j, 0 otherwise
y_i	Total working hours of nurse i
z_j	Staffing level (number of nurses required) for shift j

Table 5 Parameters

Parameter	Description
C_i	Cost associated with assigning nurse i to a shift
P_i	Preference score for nurse iii for a specific shift
W_i	Maximum working hours allowed for nurse i
S_j	Number of nurses required for shift j
T_j	Duration of shift j (e.g., 8, 12 h)
A_i	Availability of nurse i (can be binary or a specific availability window)

$$x_{ij} = \begin{cases} 1 \text{ if nurse } i \text{ is assigned to shift } j, \\ 0 \ \text{ otherwise,} \end{cases}$$

And let y_i denote the total working hours of nurse i. Using the notations in Tables 2, 3, 4, and 5, the feasible decision space X (Eq. 1) is the set of all assignment matrices $x = \{x_{ij}\}$ satisfying staffing, availability, and working hour limits:

$$X = \left\{ x_{ij} \in (0, 1)\forall i, j, \sum_i x_{ij} = S_j \forall j, x_{ij} \leq A_i \forall i, j, y_i = \sum_j T_j x_{ij}, y_i \leq W_i \forall i \right\}, \tag{1}$$

$$\text{minimize } Z_1(x) = \sum_i \sum_j C_{ij} x_{ij}, \tag{2}$$

$$\text{minimize } Z_2(x) = \sum_i y_i = \sum_i \sum_j T_j x_{ij}, \tag{3}$$

$$\begin{gathered} \text{maximize } Z_3(x) = \sum_i \sum_j P_{ij} x_{ij} \\ \text{Subject to } x \in X \end{gathered} . \tag{4}$$

This model can be expanded further with additional constraints and objectives specific to a real-world scenario, such as nurse qualifications, overtime limitations, or shift preferences. This model can be modified depending upon the type of problem we are trying to solve.

In a well-run pediatric ward, where the number of young patients stays fairly steady and the care routines are predictable, nurse managers can prepare schedules weeks in advance. Everyone knows their shifts, patients are well cared for, and the system works smoothly; this is where static nurse scheduling thrives. But walk a few steps into the emergency department, and everything changes. An ambulance rushes in, a nurse calls in sick, and patient numbers spike unexpectedly. Suddenly, the old schedule becomes useless. This is the world of dynamic nurse scheduling, where plans are rewritten on the fly, often using real-time data and intelligent systems to make rapid decisions. Nurse scheduling problems can be broadly divided into such sub-categories from stable environments like surgical or maternity wards that benefit from structured, repeatable rosters, to unpredictable, high-pressure settings like ICUs, outpatient centers, or disaster zones, where flexibility and adaptability are key. Each setting demands a different scheduling approach, not just in algorithms, but in how we think about care, resilience, and the people at the heart of the system.

Here's how you can divide these problems into following sub-categories:

Static Nurse Scheduling

1. Surgical Ward Scheduling

Surgical ward scheduling involves assigning surgeries to operating rooms, surgical teams, and post-operative beds while considering case urgency and resource availability. It aims to optimize the utilization of high-cost resources like ORs and specialized staff. The goal is to minimize delays, maximize patient throughput, and ensure safety. Emergency surgeries often require flexible, buffer capacity planning.

Objective: Minimize nurse fatigue and ensure balanced workload while maximizing coverage during surgeries.

Objective Function:

$$\min \sum_{n \in N} \text{Overtime}_n + \text{Fatigue}_n,$$

where

Overtime_n Overtime done by nurse n.
Fatigue_n Fatigue of nurse n.

Constraints:

1. Each surgery must have at least fixed number of scrub nurses and circulating nurses.
2. No nurse works more than 10 h a day.
3. Rest is mandatory after night shifts.
4. Availability of nurses must align with surgical block schedule.

2. Pediatric Ward Scheduling

Scheduling in pediatric wards requires assigning staff with specialized childcare skills to ensure safe, age-appropriate care for infants and children. Prioritizing continuity of care and emotional bonding is crucial. Schedules must also comply with stricter nurse-to-patient ratios and consider family visitation coordination. Patient vulnerability makes quality staffing essential.

Objective: Maximize care continuity and ensure child-specialized staff allocation.

Objective Function:

$$\max \sum_{p \in P} \text{Continuity}_{p,n} \times \text{Assigned}_{p,n},$$

where

$\text{Assigned}_{p,n}$ Decision variable whose value is 1 if nurse n is assigned to patient p.

$\text{Continuity}_{p,n}$ Weight in [0,1] representing preference if p is continuously cared by n.

Constraints:

1. Only pediatric nurses should be scheduled for pediatric patients.
2. A predefined nurse-to-patient ratio should be maintained throughout the schedule.
3. No more than 2 consecutive night shifts should be scheduled for any nurse.
4. Parental visiting hours must be always staffed.

3. **Maternity Ward Scheduling**

Maternity ward scheduling focuses on continuous coverage for labor, delivery, and postpartum care, involving nurses, midwives, and OB-GYNs. The unpredictability of childbirth demands round-the-clock preparedness and quick rescheduling during peaks. The goal is to ensure timely assistance, minimize fatigue, and support safe deliveries and maternal health.

Objective: Ensure uninterrupted coverage for delivery and postnatal care while minimizing overtime.

Objective Function:

$$\min \sum_{n \in N} \text{Overtime}_n + \text{Gaps},$$

where

Overtime_n Overtime done by nurse n.
Gaps Gaps between the coverage.

Constraints:

1. Labor room must be covered 24 * 7.
2. Only nurses qualified as midwives can be assigned to shifts involving childbirth/delivery.
3. No nurse should work more than 3 night shifts in a row.

4. **Nursing Homes Scheduling**

In nursing homes, scheduling ensures continuous long-term care for elderly or chronically ill patients. The emphasis is on staff continuity, emotional support, and balanced workloads. Schedules must accommodate therapy sessions, medication rounds, and personal care routines. Consistent staff assignment improves resident comfort and trust.

Objective: Maximize continuity of care and minimize nurse fatigue.

Objective Function:

$$\max \sum_{\substack{r \in R \\ n \in N}} \text{Continuity}_{r,n} \times \text{Assigned}_{r,n} - \mu \sum_{n \in N} \text{Fatigue}_n,$$

where

$\text{Continuity}_{r,n}$	Preference of continuity between nurse n and resident r.
$\text{Assigned}_{r,n}$	Binary variable to show the assignment of nurse n to resident r.
$\text{Continuity}_{r,n}$	Preference of continuity between nurse n and resident r.
Fatigue_n	Fatigue of nurse n.
μ	Trade-off between continuity and fatigue.

5. **Clinic Scheduling**

Clinic scheduling deals with organizing appointments for outpatient consultations, diagnostics, or minor procedures. It aims to reduce patient waiting time and optimize consultation room and doctor usage. Flexibility for walk-ins and urgent cases is often required. Efficient scheduling enhances patient satisfaction and staff productivity.

Objective: Minimize idle time and ensure nurse presence during all appointment blocks.

Objective Function:

$$\min \sum_{t \in T} \text{Idle time}_t,$$

where

Idle time_t Total idle time of all nurses during time block t in set T of time blocks.

Constraints:

1. Clinic hours must have complete coverage.
2. Nurse availability must be aligned with physician appointments.
3. No double-booking is allowed.
4. Lunch and break time restrictions should be there.

6. **Outpatient Center Scheduling**

Outpatient centers handle high volumes of minor surgeries, tests, and follow-ups, requiring tightly managed schedules to maintain flow. The focus is on minimizing idle time, rescheduling, and bottlenecks in shared equipment like MRI machines. Staff rotation and time slot allocation are critical to maintaining operational efficiency.

Objective: Maximize patient throughput across diagnostic/treatment rooms.

Objective Function:

$$\max \sum_{n \in N} \text{Patients Handled}_n,$$

where

Patients Handled_n No. of patients treated by nurse n.

Constraints:

1. At least one nurse per procedure room should be present.
2. There should be an upper limit for the overtime duty.
3. The schedule must allow room turnover time.

7. **ICU Scheduling**

ICU scheduling demands precise planning due to the critical nature of patient care, requiring high nurse-to-patient ratios and specialized staff. The objective is to ensure 24/7 coverage without overburdening staff. Emergencies, quick-response times, and patient acuity levels make this a high-stakes scheduling task.

Objective: Minimize nurse overload and ensure complete critical care coverage.

Objective Function:

$$\min \sum_{n \in N} \text{Workload}_n \times \text{Night Shifts}_n,$$

where

Workload_n Workload of nurse n.
Night Shifts_n Night shifts of nurse n.

Constraints:

1. Predefined nurse-to-patient ratio should be maintained.
2. Nurses should have mandatory critical care certification.
3. Buffer staff should be available in emergency.

Dynamic Nurse Scheduling

1. **Emergency Department**

Dynamic scheduling in emergency departments adapts in real-time to unpredictable patient inflows and case severity. The system must prioritize triage, minimize wait times, and deploy staff flexibly. Traditional static models fail here. AI and real-time optimization are often used to manage fluctuating workloads and ensure rapid response.

Objective: Dynamically allocate nurses to shifts to minimize patient wait and maximize triage response.

Objective Function:

$$\min \sum_{i \int E} \text{Triage Weight}_i \times \text{Wait Time}_i,$$

where

Triage Weight_i	Numeric score showing the clinical urgency of patient *i*.
Wait Time_i	The amount of time the patient *i* has to wait before being treated.

Constraints:

1. Nurses should not exceed a predefined no. of hours per week.
2. Assign only skilled nurses to specific cases like trauma or cardiac.
3. Avoid nurse burnout.

2. **Disaster and Crisis Management Scheduling**

Dynamic scheduling in disaster and crisis management involves real-time allocation of scarce healthcare resources such as ICU beds, ventilators, staff, and transport in rapidly changing, high-pressure situations like pandemics, mass casualty events, or natural disasters. The objective is to maximize patient survival, prioritize based on triage, and adapt instantly to new information, such as patient inflow or infrastructure damage. Unlike static scheduling, this requires robust, reactive algorithms such as stochastic optimization, multi-agent systems, or Reinforcement Learning to make urgent decisions under uncertainty. Coordination across multiple facilities, mobile units, and emergency services is often necessary.

Objective: To maximize coverage under uncertainty and triage response effectiveness.

Objective Function:

$$\max \sum_{i \in C} \text{Triage Score}_i \times \text{Treated}_i - \sum_{n \in N} \text{Burnout}_n,$$

where

Triage Score_i	Urgency level of patient *i*.
Treated_i	Binary variable indicating whether the patient *i* is treated or not.
Burnout_n	Fatigue or overwork score of nurse *n*.

Constraints:

1. Nurses or equipment may be shared between departments or hospitals.
2. Nurse shifts may change suddenly during a crisis.

Table 6 Using static versus dynamic nurse scheduling

Use of static nurse scheduling	Use of dynamic nurse scheduling
Patient load and nurse requirements are predictable	Sudden changes in patient volume or severity
Operations are routine and repetitive	Frequent emergencies or unpredictable events
Emergencies or disruptions are rare or minimal	Staff may be absent or redeployed
Staff availability is known and stable	Sudden post-operative complications or extended care
Care protocols and shift patterns are standardized	Need to reassign nurses in real-time

3. Always meet legal staffing levels, plus extra for emergencies.

While static nurse scheduling is commonly used in the environments mentioned above, certain sub-categories such as surgical wards, Intensive Care Units, and outpatient centers (during peak hours) may still require dynamic nurse scheduling strategies due to various factors such as unpredictable surgery durations and post-operative complications, fluctuating patient acuity levels, staff shortages and surges in patient volume. The following Table 6 gives the idea of when to use static or dynamic type of scheduling.

Benchmark Datasets

When researchers develop a new nurse scheduling algorithm, it's a bit like designing a car it might look great on paper, but until it's tested on real roads with sharp turns, traffic, and unpredictable weather, no one knows how well it truly performs. That's where benchmark datasets come in. Imagine them as carefully crafted "test tracks" that replicate the real-world complexities of hospital scheduling different shift lengths, legal staffing limits, nurse preferences, and emergency scenarios. These datasets give every new model the same conditions to prove its strength, fairness, and flexibility. Whether it's a small rural clinic or a large urban hospital, these benchmarks help ensure that scheduling tools don't just solve a math problem they support real nurses, real patients, and real care environments. In this way, benchmark datasets quietly power progress, pushing research toward solutions that actually work where it matters most: on the hospital floor. Table 7 presents the information about Standard Benchmark Datasets for Nurse Scheduling Problems.

3 Methodology for Review

On January 17, 2025, the necessary data required for the literature review and bibliometric analysis were collected from 'Web of Science' core collection database. The keywords used were ('Staff Scheduling' AND 'Nurse Scheduling') OR ('Nurse Rostering' AND 'Nurse Scheduling'). Only the papers related to staff scheduling

Table 7 Nurse scheduling datasets

Name of the dataset	Description	Features
INRC-I	• Released as part of the first International Nurse Rostering Competition in 2010 • It reflects real-life weekly rostering problems and includes constraints like skill requirements, minimum and maximum workdays, rest days, and shift sequences	• Multiple nurses and shifts • Hard and soft constraints • Historic data dependencies
INRC-II	• Extension of INRC-I by incorporating a multi-stage, long-term planning scenario • It simulates real-life monthly and weekly nurse scheduling in a more dynamic and uncertain environment	• Preferences, fairness, skill coverage • Stochastic elements (e.g., future demand uncertainty) • Planning over multiple weeks with memory of past schedules
NSPLib	• A collection of real-world and artificial nurse scheduling instances compiled by the University of Nottingham • It includes a wide range of cases used in academic literature	• Varying complexity levels • Diverse hospital sizes and shift systems • Includes both cyclic and non-cyclic scheduling
Valouxis and Housos	• One of the earliest publicly available nurse rostering datasets • Often used for evaluating basic algorithm performance	• Basic constraints (e.g., maximum consecutive shifts, rest days) • Smaller in size compared to INRC datasets
BCV	• Real-world dataset used for validating scheduling algorithms in the healthcare sector • Focuses on fairness and regulatory compliance	• Fairness in night and weekend shift distribution • Restrictions on consecutive workdays and shift sequences • Balanced workload and staffing
Ozkarahan	• One of the earliest mathematical models for cyclic nurse scheduling problems • Focuses on fairness and coverage	• Mostly used for small-scale problems
GPost	• Real-world dataset from German healthcare services • Weekly schedules with complex rules and XML format	• Complex rules and staff constraints • Includes multiple roles/skills

(continued)

Table 7 (continued)

Name of the dataset	Description	Features
SINTEF	• Designed to model real-world hospital nurse rostering problems • Particularly focuses on practical constraints from Scandinavian healthcare institutions	• Includes historical demand data • Uncertainty in availability • Predictive model validation
Ikegami-3shift data1	• Widely used in academic research to test nurse rostering algorithms under structured and simplified conditions • Developed to evaluate performance of exact methods like constraint programming and integer programming, as well as heuristic/metaheuristic algorithms	• Multiple nurses and 3 daily shifts • Uniform skill set across nurses • Restrictions on shift transitions (e.g., no Night → Morning) • Fixed scheduling horizon (28 days)
Azaiez	• Models realistic weekly scheduling scenarios in hospitals, incorporating multiple hard and soft constraints • Well-suited for both exact mathematical approaches (like goal programming, MIP) and metaheuristic techniques (GA, PSO, SA) • Particularly known for its goal programming formulation	• Focuses on: • 1. Coverage requirements • 2. Work regulations • 3. Nurse preferences
OzCare	• Real-world dataset from Australian aged-care facilities • Shared selectively in research under agreements	• Includes preferences, shift rules, seniority levels
Li et al.	• Simulated datasets based on realistic assumptions. • Up to 100 nurses over a 14-day horizon	• Multiple objectives: preferences, fairness, minimum/maximum shifts
Toronto	• One of the earliest hospital scheduling datasets	• Emphasis on fair coverage, rotation constraints
HIBISCUS	• Realistic benchmark suite from French hospitals	• Includes multiple planning instances, anonymized hospital data. • Offers tools to visualize and test constraints
Synthetic datasets	• Simulated nurse rostering instances used in genetic algorithms and metaheuristic studies	• Varying sizes, shift patterns, and constraints

in Hospital Management Systems were selected. A total of 37 papers were selected for this literature review and bibliometric analysis.

The objective of this review paper has been formulated in response to the persistent challenges in nurse scheduling, such as sudden patient arrivals, fluctuating demand, and staff shortages. Traditional scheduling methods often struggle to accommodate these dynamic factors, leading to inefficiencies and increased workload for nurses. With the growing complexity of healthcare operations, the use of AI-driven approaches in nurse scheduling has been increasing rapidly due to their ability to optimize schedules efficiently, adapt to real-time changes, and enhance decision-making. Hence, this paper aims to review recent advancements (2020–2024) in nurse scheduling, focusing on AI-based and other emerging methodologies to assess their effectiveness and practicality.

3.1 Literature Search

The nurse scheduling problem (NSP) has gained significant attention due to its critical role in ensuring efficient hospital operations, high-quality patient care, and nurse satisfaction. The growing complexity of healthcare systems has necessitated the development of innovative scheduling models. This section reviews the recent advancements in nurse scheduling research, focusing on works indexed in the Web of Science database from 2020 to 2024. The selection of research articles has been done according to several classification criteria which are shown in Fig. 4. These studies encompass a variety of approaches, including optimization techniques, heuristic methods, and scenario-based models, addressing both theoretical developments and practical applications. We have tried to provide a comprehensive foundation for future research by categorizing these works. We have also tried to highlight trends, challenges, and gaps in the field.

3.2 Related Work

Integer Programming Models for Nurse Scheduling

Integer programming (IP) has long been the cornerstone of nurse scheduling research. Its strength lies in its ability to rigorously model complex scheduling decisions with strict constraints such as legal limits on working hours, minimum staffing levels, and individualized shift preferences. Over the years, researchers have extended basic IP models into multi-stage, goal-oriented, and fairness-aware formulations, adapting them to the increasingly dynamic and human-centered demands of modern healthcare systems. The following studies highlight how integer programming continues to evolve to meet these challenges.

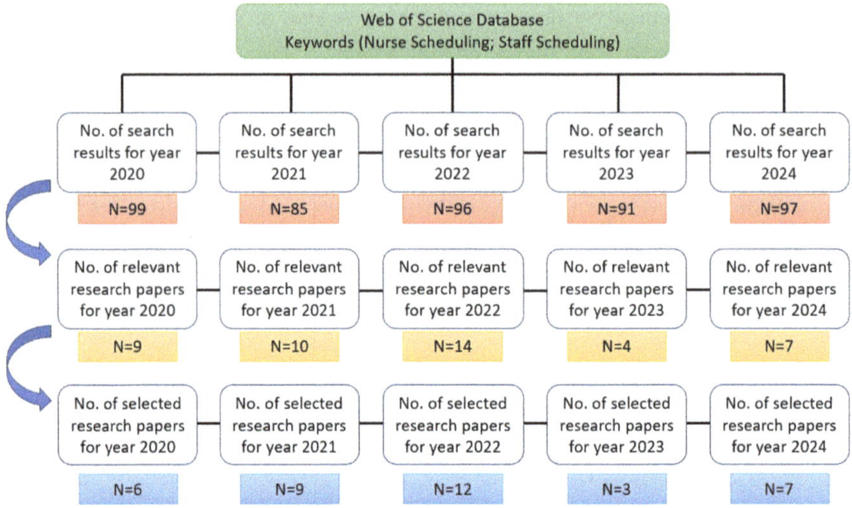

Fig. 4 Prisma diagram for selecting research articles

The journey begins with Pahlevanzadeh et al. [2], who placed fairness and reliability at the heart of scheduling. Their two-stage framework first ranks nurses using fuzzy TOPSIS and AHP factoring in teamwork, patient care quality, and documentation and then feeds those rankings into a binary IP model. The result schedules that reward strong performance while honoring preferences. By including practical constraints like kindergarten capacity and gender-specific assignments, and solving via the ε-constraint method in GAMS, they achieved a 40% reduction in absences and a 30% rise in satisfaction at Kamkar Hospital's emergency unit.

Fairness is crucial, but large hospitals need scale. Guo and Bard [5] tackled this by designing a two-stage MILP for monthly schedules covering up to 60 nurses in a 1000-bed hospital. Stage 1 used column generation to create regular-time rosters that account for birthdays, vacations, maximum consecutive work/off-days, and rest requirements with preference violations modeled as soft constraints carrying exponential penalties. Stage 2 applied a local search heuristic to minimize overtime while balancing workloads. Their approach delivered practical runtimes and robust solutions, proving that IP can handle the computational challenges of big, busy hospitals.

The next leap was the predictive power. Aydas et al. [9] explored medium-term nurse staffing policies for Intensive Care Units (ICUs) to cut costs while balancing the risks of under- or over-staffing. This study presents a medium-term integrated nurse staffing and scheduling framework for Pediatric Intensive Care Units (PICUs), using mixed-integer programming (MIP) to minimize staffing costs while accounting for under- and over-staffing risks. The authors propose two models: A Fixed Nurse Staffing Model (FNSM) using static demand estimates, and a Dynamic Nurse Staffing Model (DNSM) based on regression-forecasted non-stationary demand. The DNSM incorporates three key variables which are patient census, acuity, and admission-discharge-transfer (ADT) activity to predict required nurse hours per shift. Feasible nurse schedules are pre-generated for 30 different nurse job profiles, considering complex work rules and shift patterns. The forecast models use historical data to predict demand across a 6-week horizon, and the resulting nurse schedules are optimized for cost and service balance. Experimental results across three wards show that the DNSM consistently outperforms FNSM in reducing both cost and understaffing, especially under high understaffing penalty conditions. This integrated, data-driven approach supports more flexible, accurate staffing plans in high-demand, uncertainty-prone ICU environments.

$$\text{Min .staffing and penalty costs} = \text{FTE cost} + \text{PRN cost} + \text{Understaffing penalty},$$

where

FTE Full-time equivalent nurses.
PRN Demand-based nurses which haven't been scheduled.

Some researchers linked scheduling directly to patient experience. Pérez and Dzubay [10] developed a scheduling-based approach to enhance patient perceptions of care in ICUs, focusing on doctor communication and discharge information which were the two key areas of the HCAHPS survey. This paper introduces two integer programming models aimed at improving patient perceptions of care quality in ICUs by optimizing nurse and doctor schedules. The first model (MIP1) sequences nurse and doctor visits to ensure that both are present during ward rounds, enhancing doctor–patient communication. The second model (MIP2) minimizes the number of days doctors perform ward rounds after being on-call the previous night, thereby supporting earlier discharges. Both models address two critical dimensions of the HCAHPS survey: doctor communication and discharge information. Implemented using Excel with OpenSolver, the models were validated through simulation and real hospital data from a central Texas ICU. Results show a 30% reduction in discharge times and a 20% increase in joint nurse–doctor visits. HCAHPS scores improved significantly, doctor communication jumped from the first to the 75th percentile, and discharge information rose to the 95th percentile after one year of applying the models.

Adaptability then became the focus. Björek Böðvarsdóttir et al. [11] co-designed a modular MIP system with Danish hospitals that could flexibly incorporate ward-specific rules, soft penalties, and even novel constraints like chaperoning relationships or sleep pattern protections. Days off were treated as shift types, preferences were normalized, and exact methods (no heuristics) still produced high-quality rosters which helped cutting manual planning while improving nurse-friendliness.

$$\text{Objective Function} = \min \sum_{i=1}^{9} \Omega_i,$$

where

Ω_1 Relates to satisfaction and violation of preferences.
Ω_2 Penalizes when exceeding the preferred max. no. of consecutive days without a day-off.
Ω_3 Penalizes for deviating from targets for the assigned hours.
Ω_4 Penalizes when a series of consecutive work assignments is out of bounds.
Ω_5 Penalty for unwanted assignments.
Ω_6 Penalty for coverage related constraints.
Ω_7 Penalty for uneven distribution of excess staffing.
Ω_8 Provides a reward when trainees work together.
Ω_9 Penalty for too many work sequences.

Finally, Integer Programming entered the real-time data. Aydas et al. [12] proposed a two-stage short-term nurse schedule adjustment model that adapts staffing based on dynamic patient demand. This study proposes a two-stage stochastic integer programming model (2S-SIPM) for adjusting nurse schedules every 4 h, addressing dynamic patient demand, particularly in ICUs. The first stage involves decisions made at the start of a shift (shift t) for the next shift (shift $t+1$), including requesting or canceling nurse resources. The second stage, after demand is realized, involves corrective decisions like mandatory overtime. The model integrates historical data to estimate patient acuity and unscheduled admissions, providing more accurate short-term forecasting. Six adjustment options are modeled, five for the first stage and one for the second stage. The objective function minimizes total staffing and adjustment costs while ensuring patient coverage. The solution uses a Benders decomposition (L-shaped method) for computational efficiency, solving instances within 6 min. The model is validated using real data from a 72-bed pediatric ICU, achieving cost savings up to 18% and maintaining understaffing below 2%.

Legrain and Omer [13] proposed a branch-and-price algorithm to address personalized nurse scheduling problems involving a range of hard and soft constraints. The authors develop a branch-and-price algorithm tailored for solving a wide range of nurse scheduling problem (NSP) instances involving both hard and soft constraints. Their approach decomposes the problem using column generation where each column represents an individual nurse's schedule. The pricing problem is formulated as a shortest path problem with resource constraints (SPPRC), incorporating various scheduling rules as resource constraints, including soft bounds and forbidden patterns. Dynamic programming is used to solve the SPPRC, with specific dominance rules and resource extension functions to filter inefficient paths. The formulation captures complex personalized constraints using generic representations, allowing flexibility in modeling. They implement acceleration techniques such as strong branching, a rotation-based MIP heuristic, and parallel processing. Numerical experiments on benchmark datasets including INRC-I and INRC-II demonstrate the algorithm's ability to solve open instances optimally and improve known bounds. The code is open-source and designed for extensibility.

Hybrid Integer Programming Models for Nurse Scheduling

While pure integer programming offers precision, real hospital environments often demand more flexibility and that's where hybrid approaches shine. These methods combine exact optimization with heuristics, metaheuristics, or multi-objective balancing to handle personalization, skill matching, and conflicting goals without sacrificing computational efficiency. The studies in this category show how blending techniques can create schedules that are not only feasible, but also fair, adaptable, and scalable.

Legrain et al. [14] developed a rotation-based branch-and-price algorithm to tackle the personalized Nurse Scheduling Problem (NSP), specifically the deterministic version from the second international nurse rostering competition. This paper proposes a novel rotation-based branch-and-price algorithm to solve the nurse scheduling problem (NSP), as introduced in the second International Nurse Rostering Competition (INRC-II). Instead of constructing full rosters as in traditional models, the method generates "rotations"—partial rosters representing consecutive working days framed by rest days. These rotations are then combined to form complete schedules using an integer programming (IP) formulation. A key contribution is the pricing problem, which identifies the most cost-effective rotations using a shortest path problem with resource constraints (SPPRC). The algorithm integrates this with a large neighborhood search (ALNS) metaheuristic to improve scalability for large instances (up to 120 nurses over 8 weeks). To enhance performance, a rolling horizon heuristic generates initial solutions, while personalized constraints and skill assignments are handled in a modular fashion. Results show strong performance compared to benchmarks, confirming the efficiency of the rotation-based decomposition in balancing feasibility, personalization, and computational cost.

Where Legrain et al. focused on modular decomposition, Turhan and Bilgen [15] blended integer programming with discrete particle swarm optimization (PSO) in a meta-heuristic approach. Their model assigned nurses to shifts across multiple hospital units, considering both required and preferred skills, as well as day-on/day-off requests. Skills were modeled as both hard and soft constraints, with penalties applied for mismatches, staffing imbalances, or request violations. The hybrid workflow began with IP to produce a feasible starting schedule, followed by PSO iterations to explore better assignments. Any infeasibilities generated by PSO were repaired with targeted IP sub-models. Tested on realistic datasets, the method consistently achieved near-optimal solutions and outperformed standalone IP in larger problem instances a strong case for combining exact and swarm-based search.

$$\text{minimize } F = F_s + F_a + F_m + F_d,$$

where

F_s Accounts for shift on and off requests from nurses.
F_a Accounts for over and understaffing of nurses.

F_m Assignment of less skilled nurses when skilled nurses are preferred.
F_d Penalty situation when day-on and day-off requests are violated.

Pasandideh and Behmanesh [16] presented a mathematical model for staff scheduling in operating rooms, focusing on maximizing staff satisfaction while ensuring staffing requirements are met. Their approach uses goal programming, which outperforms current hospital practices by better optimizing staff preferences and improving scheduling outcomes. This paper presents a monthly staff scheduling model for operating rooms aimed at maximizing staff preferences while respecting hospital shift requirements. The problem is modeled using a mixed-integer linear programming (MILP) formulation enhanced with goal programming (GP) to handle multiple conflicting objectives. The model includes both hard constraints (e.g., required staff per shift, no night-to-morning transitions, specific off-days) and soft constraints (e.g., individual shift preferences, rest time, and hybrid shift combinations). The methodology incorporates a detailed classification of preferences, penalty scores, and deviation variables to measure satisfaction. A single-objective GP model is created by weighting these deviations. The solution is implemented in GAMS software and tested using real hospital data. Eight simulated weight scenarios are used to analyze performance, and ANOVA tests confirm the superiority of the proposed model over the hospital's manual scheduling approach.

$$\begin{aligned}\text{Max}.z = {} & w_{\text{VIP}}.\text{VIP} + w_{\text{OFP}}.\text{OFP} + w_{\text{ONP}}.\text{ONP} + w_{\text{EWP}}.\text{EWP} \\ & + w_{\text{NWP}}.\text{NWP} + w_{\text{DE}}.\text{DE} + w_{\text{DN}}.\text{DN} + w_{\text{EN}}.\text{EN} + w_{\text{DEN}}.\text{DEN}'\end{aligned}$$

where each term is a combination of no. different types of shifts assigned and its considered weight to form a single-objective function.

Metaheuristic Models for Nurse Scheduling

When hospital wards face unpredictable patient loads, changing staff availability, and the need to keep costs in check, traditional scheduling models can struggle. This is where metaheuristic approaches excel while drawing inspiration from nature, biology, and complex systems to find near-optimal solutions quickly, even in chaotic scenarios. The following studies show how researchers are harnessing this power to create nurse rosters that are not only efficient, but also adaptable and resilient.

Hassani and Behnamian [17] developed a sustainable approach to the Nurse Rostering Problem (NRP) using a robust, scenario-based optimization method. This paper presents a robust optimization model for the nurse rostering problem that accounts for uncertainty through a scenario-based approach combined with a pessimistic decision-making criterion. The model considers multiple realistic future demand scenarios simultaneously and uses a maximin (worst-case) objective function, aiming to generate rosters that are resilient under the most challenging conditions. The formulation minimizes a combination of costs related to nurse assignments, reserve staffing, overtime, underemployment, and preference mismatches. To solve this complex, NP-hard problem, the authors implement a discrete version of the Differential Evolution (DE) algorithm, which uses tailored mutation, crossover, and selection mechanisms. The feasibility of generated solutions is ensured through penalizing and rejection strategies for constraint violations. The DE algorithm's performance is compared to Genetic Algorithms (GA) and CPLEX, with results demonstrating that DE provides better or comparable solutions with lower computation time, especially for large, multi-scenario cases. The study emphasizes the value of robustness in healthcare scheduling, especially in crisis settings.

$$\text{Objective Function} = \text{Minimize } A + B + C + D + E,$$

where

- A Min. nurse allocation cost.
- B Reserved nurses cost.
- C Overtime cost.
- D Underemployment cost.
- E Preference mismatch cost.

To focus on accessibility and modernization, Leung et al. [18] explored the use of cloud computing and mobile devices for optimizing nursing staff demand modeling and scheduling in nursing homes. This paper presents an integrated, cloud-based multi-objective optimization framework for modeling nursing staff demand and scheduling in government-subvented nursing homes. The model addresses the challenge of balancing cost, overtime, and nurse-to-resident ratios, while accounting for skill mix, heterogeneous workforce, and resident case mix. The authors formulate a multi-objective mathematical model with three goals: minimize total staffing cost, minimize overtime, and maximize staffing level. The model is solved using NSGA-II, a nondominated sorting genetic algorithm, embedded within a cloud computing infrastructure to enhance accessibility and efficiency. Chromosomes encode shift and task assignments, and feasibility is ensured through strict constraints. A real-world case study in a Hong Kong nursing home demonstrates improved staffing efficiency, reduced cost, and higher nurse-to-resident ratios. The mobile application interface allows nursing staff and managers to retrieve optimized schedules in real-time, showing the practical benefit of digitalization in elderly care planning.

Objective function 1 = Minimize total staffing costs of all nursing staff.
Objective function 2 = Minimize total overtime work of regular nursing staff.
Objective function 3 = Maximize staffing levels.

Finally, Sarkar et al. [19] took inspiration from the animal kingdom to handle dynamic scheduling challenges. Their hybrid Sea Lion Attacking–based Deer Hunting Optimization Algorithm (SLA-DHOA) first generated a library of static schedules stored in the cloud. Then, week by week, it adapted those schedules to changing demand by blending the global search power of Sea Lion Optimization with the precise local improvements of Deer Hunting Optimization. This two-phase design allowed the algorithm to avoid local optima while converging quickly to high-quality solutions. Tested against popular algorithms like PSO, GWO, WOA, and RDA in MATLAB, SLA-DHOA delivered superior feasibility, speed, and accuracy, showing clear potential for real-time hospital scheduling.

Hybrid Metaheuristic Models for Nurse Scheduling

In nurse scheduling, no single approach has all the answers. The complexity of real-world constraints, human preferences, and shifting hospital demands has inspired researchers to merge the strengths of multiple optimization paradigms, i.e., creating hybrid metaheuristics that can think, adapt, and learn.

Zhang et al. [4] introduced MDQN-MA, a pioneering fusion of deep Reinforcement Learning and multi-agent systems. Imagine four "virtual schedulers," each with its own personality, exploring the schedule space in parallel. Guided by Deep Q-Networks, these agents learn over time which local changes yield the biggest long-term improvements. Their solutions are pooled and the best one moves forward, aided by a DQN-based Tabu Search with experience replay to avoid getting stuck. Tested on INRC-II benchmarks, this multi-agent learning framework beat state-of-the-art metaheuristics showing how AI can coordinate diverse search strategies like a team of experts brainstorming together.

Chen and Zeng [20] took a different hybrid route of blending decision tree logic and greedy search with the Bat Algorithm and Particle Swarm Optimization. The decision tree, trained on domain-specific rules, generates feasible schedules from the start by avoiding hard constraint violations. Then the greedy search polishes the results by minimizing soft constraint penalties weighted through Analytic Hierarchy Process. This "smart start + local fine-tuning" approach greatly improved convergence and solution quality in a real-world radiology department case study. Below given is the objective function which has been formulated in the above paper.

$$\text{Min. Penalty} = \sum_{\substack{i,j \in \{1,2,3\} \\ \text{and } i<j}} \sum_{n=1}^{N} \sum_{d}^{D-1} W_{ij} \times Pij_{n,d},$$
$$+ \sum_{i=1}^{3} \sum_{n=1}^{N} \sum_{d}^{D-2} W_{RiR} \times PRiR_{n,d} + \sum_{n=1}^{N} \sum_{d=1}^{D-2} W_{3R1} \times P3R1_{n,d}$$

where

W_{ij} Penalty for a nurse working shift *i* followed by shift *j* the next day.
W_{RiR} Penalty for a nurse working patterns of 'off-shift *i*-off' on 3 consecutive days.
W_{3R1} Penalty for a nurse working patterns of shift3-off-shift1.

Others are binary variables taking value 0 or 1.

When COVID-19 stretched healthcare to its limits, Abadi et al. [21] responded with HSSAGA, a hybrid of Salp Swarm Algorithm and Genetic Algorithm. Their model allocated nurses across ICU and non-ICU wards in Iran, balancing expertise, workload, infection exposure, and rest days. The hybrid design leveraged SSA's exploration and GA's recombination strengths, outperforming other evolutionary algorithms in both speed and quality. The result: better rotation planning, reduced clustering of ICU exposure, and more rest for exhausted nurses which was critical for both patient safety and staff well-being.

Muniyan et al. [22] proposed an artificial bee colony algorithm integrated with the Nelder–Mead method (NM-ABC) to solve the Nurse Scheduling Problem (NSP). This paper introduces a hybrid metaheuristic algorithm combining the Artificial Bee Colony (ABC) algorithm with the Nelder–Mead (NM) method to solve the Nurse Scheduling Problem (NSP). The core idea is to enhance local exploitation in the ABC algorithm by applying the NM method during the onlooker bee phase, allowing higher-quality solutions to converge faster. In this paper, the objective function is purely based upon minimizing preference penalties. The algorithm handles both hard constraints (e.g., coverage requirements, legal work limits) and soft constraints (e.g., preferences, shift succession) in a multi-objective formulation, with penalty-based evaluation. The hybrid NM-ABC was tested on NSPLib datasets, including instances with up to 60 nurses across 28 days and 4 shifts. Performance was benchmarked against five algorithms (MAPA, HABC, BCO, HEAS, HSHH), showing superior results in terms of cost minimization, feasibility rate, and convergence time. The NM-ABC achieved best results in 44 out of 75 instances, and reduced cost gap by up to 94.3% compared to known optima.

AI-driven hybrids are also on the rise. Chen et al. [23] propose a neural network-assisted method to solve the nurse rostering problem (NRP), aiming to optimize nurse scheduling while considering individual preferences and constraints. This study proposes a hybrid approach integrating deep learning into nurse rostering to automate heuristic design and improve solution quality. The core method combines a deep neural network (DNN) to guide heuristic selection and a recurrent neural network (RNN) for reconstructing parts of schedules to escape local optima. Schedules are treated as matrices, and DNNs classify which heuristic (out of four) should be applied next. When the DNN-selected heuristic leads to stagnation, RNN reconstructs parts of the solution by generating new shift sequences based on learned patterns. A probability-penalty hybrid strategy is used for heuristic selection, balancing exploration and exploitation. The algorithm is trained using benchmark solutions and tested on standard instances, outperforming several state-of-the-art metaheuristics like CP-ILS and DP-VNS. Experiments confirm the method's competitive performance, especially in larger instances where exact methods struggle.

$$\text{Min}.\sum_{n\in N}\sum_{t\in T}\sum_{s\in S} q_{nts}(1-x_{nts}) + \sum_{n\in N}\sum_{t\in T}\sum_{s\int S} p_{nts}.x_{nts} + \sum_{t\in T}\sum_{s\in S} y_{ts}.v_{ts}^{\min}$$
$$+\sum_{t\in T}\sum_{s\in S} z_{ts}.v_{ts}^{\max},$$

where

q_{nts} Penalty if shift type s is not assigned to nurse n on day t.

x_{nts} 1 if nurse n is assigned shift type s on day t; 0 otherwise.
p_{nts} Penalty if shift type s is assigned to nurse n on day t.
y_{ts} Total below the preferred cover for shift type s on day t.
$v_{ts}^{\min}$ Weight if below preferred cover for shift type s on day t.
z_{ts} Total above the preferred cover for shift type s on day t.
$v_{ts}^{\max}$ Weight if exceeding the preferred cover for shift type s on day t.

Chen et al. [24] propose a neural network-assisted approach to solve the nurse rostering problem (NRP), a complex NP-hard scheduling task involving hard and soft constraints. Their method integrates two neural models: a deep neural network (DNN) and a recurrent neural network (RNN). The DNN is trained to guide heuristic selection during local search, choosing from four heuristic strategies. The RNN is embedded in a reconstruction mechanism to modify parts of a solution and help escape local optima by generating feasible shift sequences for selected nurses. Both networks are trained offline on benchmark datasets using supervised learning. The scheduling problem is represented as a matrix to facilitate input into the networks. During optimization, the method iteratively alternates between DNN-guided heuristic improvement and RNN-based reconstruction until a stopping criterion is met. The DNN uses a hybrid selection metric combining heuristic confidence and solution penalty. Experiments validate the method's performance across various benchmark instances, showing competitiveness against state-of-the-art algorithms.

Harb et al. [25] proposed an intelligent optimization strategy for nurse–patient scheduling in Internet of Medical Things (IoMT) applications. The authors propose a Nurse–Patient Scheduling (NPS) mechanism that integrates biosensor data and hybrid optimization to address scheduling challenges in hospitals. The methodology involves three main phases: patient monitoring, patient classification, and nurse scheduling. In the monitoring phase, biosensors collect vital signs like blood pressure and heart rate at regular intervals. The classification phase uses shapelet-based pattern recognition and Dynamic Time Warping (DTW) to assign severity levels (low, medium, high) based on physiological data. The final phase introduces a hybrid Genetic Algorithm–Particle Swarm Optimization (GA-PSO) technique to generate nurse schedules that balance workloads and prioritize patient criticality. The fitness function incorporates distances between patients, their priorities, and nurse capacity. Simulation tests were conducted using real patient data and varied nurse–patient configurations, demonstrating the approach's efficiency, fairness in workload distribution, and energy savings at sensor level.

Reinforcement Learning Models for Nurse Scheduling

In a busy hospital, schedules rarely go according to plan. A sudden nurse absence, an unexpected patient surge, or an emergency ward opening can throw even the most carefully crafted rosters into chaos. Traditional scheduling models struggle in these moments but Reinforcement Learning (RL) offers a way to adapt on the fly. By learning from past adjustments, RL can make quick, intelligent changes that keep the system running smoothly without overburdening staff. Reinforcement Learning (RL) offers a dynamic solution by adapting schedules in real-time. Unlike traditional methods, RL learns from past adjustments, improving efficiency and minimizing disruptions caused by sudden absences.

Nagayoshi et al. [26] presented a dynamic nurse scheduling approach using Reinforcement Learning (RL) to handle sudden nurse absences while maintaining schedule feasibility. Initially, a constructive nurse scheduling system generates shift schedules, which are then adjusted using Q-learning to minimize rule violations when a nurse is unexpectedly absent. The RL agent selects optimal shift exchanges based on state variables like previous exchange day, total violations, and shift type. Computational experiments with 23 nurses across three shifts (day, evening, late-night) showed that the system successfully adjusted schedules in most cases, except when absences occurred late in the month with no available replacements. Results demonstrated that RL effectively modified schedules while balancing workloads, making it a promising approach for real-time nurse rescheduling. Future work aims to improve handling of critical shift absences and refine scheduling constraints.

Muklason et al. [27] discussed solving the Nurse Rostering Problem using Reinforcement Learning combined with a Simulated Annealing with Reheating Hyper-heuristic Algorithm. The study focuses on improving nurse scheduling efficiency while balancing constraints such as shift preferences, workload distribution, and hospital regulations. The proposed method enhances solution quality by leveraging Reinforcement Learning for adaptive decision-making. The results demonstrate that this approach outperforms traditional heuristic methods. Traditional scheduling methods often fail to handle the complexity and dynamic nature of the problem, leading to inefficiencies. Computational techniques, including meta-heuristics and hyper-heuristics, have been widely used to automate scheduling and improve efficiency. Reinforcement Learning (RL) is emerging as a powerful tool in this domain, as it learns from past scheduling decisions and adapts to changing conditions. By integrating RL with optimization techniques like simulated annealing, scheduling systems can explore better solutions while avoiding local optima. Simulated annealing with reheating helps overcome stagnation in optimization by periodically increasing the search space exploration. This approach ensures that nurse schedules meet hospital requirements while balancing workload distribution. RL-based nurse rostering methods can significantly reduce scheduling conflicts, improve nurse satisfaction, and optimize hospital operations. Additionally, these methods help minimize the time required for manual adjustments, allowing hospital administrators to focus on other critical tasks. As nurse scheduling remains an NP-hard problem, RL-based techniques continue to offer promising advancements in real-time and large-scale scheduling applications.

Together, these RL-based approaches point toward a future where hospital schedules don't just react to problems but they anticipate them, learning and improving with every shift change. In such a system, the timetable isn't a static document on a bulletin board; it's a living, learning entity that evolves alongside the hospital's needs.

Search Methods for Nurse Scheduling

Finding the perfect nurse roster can feel like solving a puzzle with hundreds of moving pieces. Even when you get it right, one wrong move like a shift conflict, a shortage, or an overbooked nurse can throw the whole thing off. That's where search-based algorithms come in: They explore countless possibilities, reshuffle schedules, and zero in on solutions that balance hospital demand with nurse well-being.

Abdelghany et al. [28] approached the problem with a two-stage Variable Neighborhood Search (VNS) that thinks strategically about penalties. In the first stage, it hunts down the costliest scheduling mistakes particularly nurse shortages using two custom neighborhood structures designed to fix the worst problems first. The second stage switches focus to fine-tuning, minimizing overall penalties by swapping shifts vertically and horizontally. The process begins with a greedy heuristic to create a decent starting roster, which is then refined iteratively. Tested on 24 benchmark datasets from Curtois and Qu, this method beat the ejection chain heuristic in most cases, showing especially strong results in medium-sized problems. For larger rosters, giving the algorithm more time yielded even better schedules is a proof that this "biggest problems first" strategy pays off.

Goh et al. [29] addressed the nurse rostering problem using a two-stage approach. This paper proposes a novel two-stage metaheuristic framework for solving the NP-hard nurse rostering problem. In Stage 1, a feasible solution is generated using a hybrid of Monte Carlo Tree Search (MCTS) and Hill Climbing (HC). A customized C-value decay strategy and a new tree selection policy improve MCTS performance by balancing exploration and exploitation. If MCTS fails to find feasibility, HC refines the partial solutions. In Stage 2, the feasible roster is optimized using Iterated Local Search (ILS) with Variable Neighborhood Descent (VND) as its core. Three neighborhood structures namely horizontal swap, vertical swap, and extraction, i.e., enable flexible moves, while a perturbation mechanism helps escape local optima. The algorithm outperforms several state-of-the-art solvers across 24 benchmark instances, especially in large-scale cases, where it finds new best results in 10- and 60-min runs. Its strength lies in quickly achieving feasibility, allowing more time for high-quality optimization.

$$\begin{aligned}\text{Obj.fn} = \text{Min}.\Bigg(&\sum_{n\in N}\sum_{d\in D} q_{nds} - (1 - x_{nds}) + \sum_{n\in N}\sum_{d\in D} p_{nds}\cdot(x_{nds}) \oplus \\ &+ \sum_{d\in D} w_{ds,\min}\cdot(y_{ds}) + \sum_{d\in D} w_{ds,\max}\cdot(z_{nds})\Bigg),\end{aligned}$$

where these terms are a combination of "Shift on penalty," "Shift off penalty," and "Date specific cover penalty."

Chen et al. [30] developed modified Variable Neighborhood Search (VNS) approaches to solve the nurse scheduling problem (NSP). The authors propose seven modified Variable Neighborhood Search (VNS) algorithms to solve the nurse scheduling problem (NSP), incorporating both hospital/government constraints and nurse preferences. The methodology introduces three greedy-neighborhood-swapping mechanisms—greedy-2-exchange, greedy-3-exchange, and greedy-4-exchange—which identify nurses with the highest soft constraint violations and swap their shifts to improve schedule quality. These mechanisms are used in one-, two-, and three-neighborhood configurations to form seven distinct VNS variants. A shaking phase generates diverse neighbor solutions, followed by local search using a Variable Neighborhood Descent (VND) method. Recovery schemes are implemented to ensure feasibility by correcting hard constraint violations. A mutation strategy is introduced to escape local optima by occasionally regenerating new feasible solutions. Experiments are based on real hospital data from a radiology department and evaluate solutions on preference violation penalties. Results show the greedy-4-exchange and greedy-2-exchange combination yields the best trade-off in solution quality and computation time.

Review Articles in Nurse Scheduling

Sometimes, to see where a field is going, you first have to look at where it's been. Nurse scheduling research has evolved through countless algorithms, datasets, and real-world experiments, and review studies act as the map by showing the trends, the turning points, and the roads yet to be explored.

Ngoo et al. [6] took on the ambitious task of charting nearly a decade of progress in the Nurse Rostering Problem (NRP), from 2012 to 2021. Their survey covered more than 50 solution approaches, neatly organizing them into categories: heuristics, meta-heuristics, hyper-heuristics, mathematical optimization, matheuristics, and hybrid models. They also compared the most widely used benchmark datasets—INRC-I, INRC-II, ORTEC, and NSPLib—and found that, after 2018, the field began leaning heavily toward matheuristics and hybrid approaches, often powered by population-based metaheuristics. The review didn't just stop at what's popular; it examined how factors like constraint complexity, search space size, and solver choice (CPLEX, Gurobi, etc.) influence a method's success. Scalability, adaptability to different hospitals, and handling uncertainty emerged as persistent challenges. Looking forward, the authors see opportunities in hyper-heuristics, fuzzy logic models, and intelligent, data-driven scheduling systems.

While Ngoo et al. mapped the global landscape, Little and Choudhury [7] zoomed in on Canada's healthcare system. Their review didn't just focus on nurses but it also looked at physicians, operating rooms, and outpatient scheduling, all bound by the same headaches: conflicts, time constraints, and human factors. They classified scheduling problems into deterministic static and stochastic dynamic types and then dissected over 40 real-world implementations in Canadian hospitals. Their analysis showed a mix of techniques from mixed-integer programming and constraint programming to metaheuristics like tabu search and genetic algorithms which were often customized for specific hospital workflows. Recently, they noted, the conversation has shifted toward machine learning for demand forecasting, patient load prediction, and collaborative scheduling. Yet, a recurring theme was the gap between models that work on paper and those actually used in practice.

Then came an unexpected and urgent case study: the COVID-19 outbreak. Zhang et al. [8] conducted a scoping review of nurse scheduling strategies used by medical teams deployed to Hubei Province in early 2020. They found that during the crisis, scheduling wasn't just about efficiency, but it was about survival, both for patients and for staff. Across 17 studies from January to October 2020, four main shift lengths emerged: 2, 3, 4, and 6 h, with 6 h shifts becoming the norm in makeshift hospitals for mild cases. Scheduling models ranged from dynamic workload-based approaches to humanized schedules prioritizing nurse well-being, and professional-integrated models that matched expertise with patient needs. The review offers a rare glimpse into how nurse scheduling adapts under extreme pressure, where flexibility, empathy, and adaptability outweigh mathematical elegance.

Together, these reviews don't just summarize but they tell a story of a field that is constantly evolving, often shaped by crises, technology shifts, and the human realities of healthcare work.

Other Models in Nurse Scheduling

Not every nurse scheduling problem fits neatly into standard optimization, meta-heuristics, or Reinforcement Learning categories. Some situations demand creative, domain-specific approaches from real-time crisis response to infection control, patient monitoring, and even ergonomic workload balancing. The following studies showcase how researchers have adapted scheduling models to highly specialized needs.

Schoenfelder et al. [1] developed a nurse scheduling model that uses quick-response strategies like reassigning cross-trained float nurses and transferring patients between units or off-unit admissions. This study proposes a multi-stage stochastic programming model that integrates nurse scheduling with two classes of quick-response methods: dynamic reallocation of cross-trained float nurses and patient transfers or off-unit admissions. The methodology begins with the generation of an initial schedule based on nurse availability and preferences. Patient demand is modeled using multiple stochastic scenarios to reflect real-world uncertainty. The model anticipates demand variability and incorporates real-time recourse actions at multiple decision stages. A Full Foresight policy accounts for all demand scenarios during scheduling, while a No Foresight policy assumes average demand and makes reactive adjustments. The model is validated using real data from three U.S. hospitals, differing in staffing policies. By embedding flexibility into the initial scheduling process, the Full Foresight method significantly reduces penalties from understaffing and patient turnaways. The study shows that a small number of strategic reassignments and transfers are sufficient to handle variability efficiently, highlighting the operational value of stochastic programming combined with real-time quick-response mechanisms.

Saad et al. [3] propose a three-phase framework to enhance nurse–patient scheduling in healthcare environments using wireless body sensor networks (WBSNs). The first phase involves real-time data collection and transmission, utilizing an Early Warning Score-based method to reduce energy consumption by only transmitting data when patient condition changes. The second phase focuses on patient classification through machine learning, using shapelet-based pattern recognition and dynamic time warping to assess the severity of patient vital signs and classify them as low, medium, or high risk. The third phase introduces two algorithms—PDCB and PSG—for assigning patients to nurses in balanced clusters based on criticality and number. It then applies a Patient-Nurse Scheduling (PNS) algorithm that considers both patient criticality and age to prioritize and schedule visits. Simulation is based on real sensor data from 18 patients and demonstrates improvements in workload balance and up to 95% reduction in sensor data transmission. The methodology ensures adaptive scheduling that is energy-efficient and clinically responsive.

Legrain et al. [31] developed an online stochastic algorithm to tackle the dynamic Nurse Scheduling Problem under uncertainty, as featured in the second international nurse rostering competition. This paper presents an online stochastic algorithm designed to solve the dynamic nurse scheduling problem under uncertainty, as posed in the INRC-II competition. The key challenge is to compute personalized weekly nurse rosters sequentially, without knowledge of future demand or preferences, while maintaining global feasibility and near-optimality over a multi-week horizon. The authors combine elements of primal-dual algorithms (to make locally optimal irrevocable decisions) and Sample Average Approximation (SAA) (to evaluate decisions over multiple future scenarios). The approach generates candidate weekly schedules using a static branch-and-price solver, then ranks and selects the best schedule based on sampled future demand scenarios. To address shift succession and workload balancing across weeks, the method anticipates border effects by solving two-week subproblems and aggregates decisions using contract-based soft constraints. Results show strong performance on benchmark instances, demonstrating the algorithm's robustness, scalability, and ability to adapt in real-time with minimal look-ahead data.

D'souza et al. [32] developed an automated nurse scheduling system named ROTA for a multi-specialty tertiary teaching hospital in South India. This paper introduces ROTA, an automated nurse scheduling system developed for a 2032-bed tertiary hospital in South India with 1800 nurses and 98 wards. The methodology involved three stages: exploratory discussions with stakeholders, business and software requirement gathering, and system design. The system was built using Java Spring Framework, MySQL, and cloud deployment, and it automates daily, weekly, and monthly scheduling. ROTA handles complex constraints such as nurse-patient ratios, skill levels, legal regulations, and personal preferences. Key features include dashboard views, face sheets, leave analytics, shift swaps, and training allocation. The system reduced nurse scheduling time by 78%, saved over ₹2.3 million annually, and improved managerial oversight. A satisfaction survey ($n = 210$) showed 91% nurse satisfaction, highlighting ROTA's effectiveness in improving transparency, reducing manual workload, and optimizing staff distribution across departments.

Valdano et al. [33] proposed a strategy to reorganize nurse schedules in a short-stay geriatric ward to reduce healthcare-associated infection (HAI) risks. This paper introduces a novel infection control strategy that reorganizes nurse schedules using contact network data and a Susceptible-Colonized-Susceptible (SCS) transmission model. Using RFID-based contact tracing data from a geriatric hospital ward in France, the authors propose an optimization-based intervention that reassigns tasks between nurses by modifying their work shifts while preserving the original contact structure and quality of care. A potential function minimization approach is used to generate new shift arrangements under different feasibility constraints (e.g., maximum shift length, workload balance). Four scheduling models ($S \pm 1$ and WS $\pm$ 1) are compared. The model $S + 1$, which promotes irregular but balanced schedules, achieved the greatest reduction in infection risk—27%, without removing any contacts or compromising operations. Compared to interventions like isolation or contact removal, this method is less disruptive and more sustainable. The results highlight how shift regularity, workload equity, and contact distribution influence the spread of HAIs and suggest that schedule design can be a powerful, underused infection control tool.

Cao et al. [34] introduced an efficient constructive method to determine the minimum number of staffs needed for effective scheduling, addressing a crucial step before creating work schedules. This paper proposes a constructive scheduling method that efficiently calculates the minimum number of staffs required to meet demand under various constraints. Unlike traditional methods that assume a fixed workforce size, this approach begins by determining whether the available manpower is even sufficient to create a feasible schedule. The method addresses three problems: (1) a basic model with days-off requirements, (2) a version with days off within short periods, and (3) a nurse rostering variant with multi-shift assignments. The algorithm builds a schedule matrix using matrix filling logic, where each worker is assigned workdays top-to-bottom and left-to-right to fulfill daily demands while respecting rest periods. The model runs in linear time, outperforming methods like integer programming and genetic algorithms in speed and feasibility assurance. The derived minimum staff count can also serve as a sentinel—if a problem has fewer workers than this bound, no feasible schedule exists. The method was tested against genetic algorithms and demonstrated superior performance and solution quality.

Akin and Ordu [35] proposed a simulation-based two-stage optimization approach for nurse planning to address the dynamic needs of healthcare services, particularly during the COVID-19 pandemic. This study proposes a simulation–optimization framework to determine weekly nurse requirements and generate fair shift schedules in COVID-19 inpatient services. The first step involves a discrete event simulation model built using Simul8 to capture nurse demand under varying pandemic scenarios. Four case studies and four dynamic scenarios are tested based on changing patient flow, length of stay, and ICU transfers. Outputs from simulation feed into a two-stage mathematical optimization model. The first stage computes weekly shift allocations, minimizing imbalances and satisfying legal constraints. The second stage generates daily nurse schedules, considering fairness, non-consecutive shifts, and weekend balancing. Both models are implemented in GAMS and solved rapidly, showing excellent performance on real hospital data. This approach enables proactive, data-driven planning under uncertainty, and is adaptable to future pandemics or demand surges.

Taiwo et al. [36] examined the appointment scheduling problem in physician-nurse teams, considering factors like patient choice and no-shows. The authors develop a mathematical model for appointment scheduling in clinics with both physicians and nurse practitioners (NPs), incorporating patient choice of provider. The model assumes scheduled patients choose their provider upon arrival, based on waiting time and perceived care quality. Patients are assumed to incur linear waiting costs, and their provider preference depends on a utility difference threshold. The model incorporates stochastic patient arrivals with slot-dependent no-show probabilities and models queue dynamics based on patient decisions. It explicitly accounts for provider-specific idleness and overtime. Due to the model's structural complexity and non-multimodularity, the authors derive sufficient conditions for tractability and propose a heuristic (Expected Utilization Criterion) that schedules patients to maximize provider utilization. The heuristic is proven optimal when no-shows are absent. Numerical experiments validate the approach, comparing it against "standalone" and "ice-breaker" models, and assess trade-offs in patient waiting, provider utilization, and clinic profitability.

Ding's [37] observational study evaluates the impact of a comprehensive collaborative scheduling system combined with a quantitative scoring appraisal method for operating room nurses in a primary hospital. Thirty nurses participated in the study from January to December 2022. The new scheduling model replaced rigid time blocks with flexible shifts based on daily surgical demands, and incorporated rules for compensatory time and overtime pay. The performance appraisal assigned points for various tasks, including hand-washing, post-op care, emergency response, and compliance with protocols, while deducting points for errors or negligence. Metrics such as surgical volume, working hours, and satisfaction rates were compared to a traditional scheduling and average-distribution performance model. Statistical analysis used SPSS with t-tests and chi-square tests. The new system led to higher surgical throughput, fewer average daily working hours, and improved satisfaction among nurses, doctors, and patients. This method shows promise for improving operational efficiency and stakeholder satisfaction in small hospitals.

Durak and Mutlu [38] developed a mathematical model to address the Home Health Care Routing and Scheduling Problem (HHCRSP) with a novel integration of ergonomic risk considerations. The model accounts for typical real-world constraints such as nurse skills, patient time windows, and mandatory lunch breaks, while introducing subjective workload constraints based on perceived ergonomic risks. A Fuzzy Inference System (FIS) is used to quantify nurse workloads from post-task questionnaires addressing physical, environmental, and mental ergonomic factors. These fuzzy workload scores feed into the optimization model, which aims to maximize patient visits and minimize nurse overload. The model includes constraints to ensure feasibility, such as time limits and workload thresholds. A comparison between models with and without workload constraints shows that including ergonomic risks reduces average workload and improves workload balance. Numerical experiments and sensitivity analyses confirm the robustness and practical relevance of the approach in real-world HHC settings.

Booker et al. [39] examined nurse rostering practices in Australia, focusing on how they affect staff sleep, well-being, and fatigue. This study uses a qualitative, instrumental single case study design to explore current nurse rostering practices in three hospitals in Victoria, Australia. Twenty-four nurses responsible for creating rosters were interviewed through semi-structured interviews. Data were collected remotely and analyzed using selected grounded theory and thematic analysis to identify patterns and develop categories. The study investigates who performs rostering, the extent of their training, and their awareness of fatigue management policies. Most participants had no formal training and relied on informal handovers. The dominant rostering method was self-rostering, where staff submit preferred shifts, often prioritized over fatigue considerations. Participants varied in their perception of fatigue responsibility, with many attributing it to individual nurses. The analysis also highlights barriers to change, including workplace culture and resistance to altering established shift preferences.

Table 8 gives us the comparative information about no. of constraints, no. of objectives, no. of variables, no. of parameters, tools/software used and type of optimization (single objective optimization/multi-objective optimization) considered in each paper considered for the review.

3.3 A Few Case Studies in Available Literature

A Case Study in Egypt

Adoly et al. [40] introduced a mathematical model for the Nurse Scheduling Problem (NSP), which was designed using the concept of a multi-commodity network flow framework. The model was tested on both hypothetical scenarios and standard benchmark datasets, and was further validated through application to a real-world case study from a hospital in Egypt. They have accounted for the cost incurred per shift for each nurse, the additional cost for nurse overtime per shift, and the shift-wise cost associated with assigning a head nurse. The model was prepared to generate the schedule for one week and it was tested on 2 different instances. The instances were solved using Gurobi 7.0.1 optimization software, which employs the branch and bound algorithm. All computational experiments were conducted on a system equipped with an Intel Core i7-6500U processor (2.50 GHz) and 8 GB of RAM. The results show that the proposed model increased the fairness in terms of working hours in the distribution of the shifts. Also, there was a reduction in the total cost by 36% and 31% respectively. The time needed to generate weekly schedules was negligible which meant that the computation time was less than a second.

Table 8 Comparative information of the research articles

S. No.	Year	References	No. of const.	No. of obj.	No. of var.	No. of para.	Tools/ software	Type of optimization
1	2020	[1]	15	2	7	18	CPLEX + Gurobi	MOO
2	2020	[2]	9	1	5	10	IBM ILOG CPLEX	SOO
3	2020	[3]	8	1	4	14	IBM ILOG CPLEX	SOO
4	2020	[4]	13	1	8	13	Classic DE	SOO
5	2020	[5]	11	1	4	5	C++	SOO
6	2020	[6]	11	1	4	5	CPLEX/ Gurobi	SOO
7	2021	[7]	23	2	3	13	GAMS	MOO
8	2021	[10]	13	1	4	17	MATLAB	SOO
9	2021	[11]	15	2	6	9	COIN/ CBC	MOO
10	2021	[12]	44	1	31	45	Gurobi	SOO
11	2021	[14]	29	3	16	14	–	MOO
12	2021	[15]	6	1	7	8	–	SOO
13	2022	[16]	11	1	4	20	Gurobi	SOO
14	2022	[18]	6	1	–	–	MATLAB	SOO
15	2022	[20]	9	1	2	5	MATLAB	SOO
16	2022	[21]	6	1	6	24	IBM ILOG CPLEX	SOO
17	2022	[22]	13	1	5	15	IBM ILOG CPLEX	SOO
18	2022	[23]	23	1	6	20	CPLEX	SOO
19	2022	[24]	10	1	4	20	Python	SOO
20	2022	[25]	35	1	13	10	GAMS	SOO
21	2022	[27]	3	1	5	4	GAMS	SOO
22	2023	[28]	10	1	4	15	Python	SOO
23	2024	[31]	3	1	2	–	Python	SOO
24	2024	[32]	5	1	3	8+	C++	SOO
25	2024	[33]	12	1	1	8	Python	SOO

(continued)

Table 8 (continued)

S. No.	Year	References	No. of const.	No. of obj.	No. of var.	No. of para.	Tools/ software	Type of optimization
26	2024	[35]	14	1	9	10	ILOG CPLEX + Python	SOO
27	2024	[36]	17	1	5	10	ILOG CPLEX	SOO

A Case Study in Kuwait

A case study by M'Hallah et al. [41] in Kuwait has prepared Nurse Schedule for a health care unit using an Integer Programming Model. They have considered 3 types of shifts in a day. Each nurse has been characterized by her religion, native language, citizenship, and gender. Schedule has been prepared for certain number of weeks. The objective was to minimize the number of outsourced nurses during this period while satisfying all the constraints. The case study focuses on nurse scheduling at Al-Saqer Health Center in Kuwait. The objective was to minimize the number of outsourced nurses while satisfying various operational constraints, including minimum staffing levels by gender and nationality, weekly working hour limits (42–48 h), mandatory rest days, and specific rules for head nurses and nurses in training. The model also incorporated nurses' personal requests for shifts or days off, which were previously ignored in manual scheduling. Using CPLEX via GAMS, the model produced a feasible and near-optimal schedule that significantly outperformed the manually prepared timetable in terms of constraint satisfaction and outsourcing requirements, only one outsourced nurse was needed compared to over a hundred shifts in the manual plan. The optimized schedule ensured equitable workload distribution and compliance with regulations, enhancing both nurse satisfaction and patient safety. Additionally, the model allowed sensitivity analysis by simulating nurse reassignment scenarios. This case highlights the effectiveness and practicality of Operations Research methods in solving complex healthcare scheduling problems.

A Case Study in France

This case study by Yasmin et al. [42] focuses on assigning work shifts to nurses within a hospital unit over a defined planning horizon, aiming to generate timetables that satisfy staffing requirements while adhering to contractual rules and individual working conditions. The objective function integrated multiple goals, including minimizing violations of soft constraints such as rest periods and shift preferences, and ensuring balanced workloads, equitable distribution of unpopular shifts, and fair assignment of weekends and holidays. The model was tested on cases ranging from 10 to 40 nurses and 4 to 8 shifts—reflecting common ward scenarios. A flexible MILP-based scheduling model was developed and implemented in a French hospital, incorporated into a decision support system to enhance usability. The model simultaneously addressed staff well-being, equity, and operational efficiency. Compared to manually created schedules, the algorithm improved workload equity by over 50%, night-shift distribution by over 87%, and soft constraint satisfaction by 47%. Additionally, it drastically reduced scheduling time from several hours or days to just minutes. Overall, the proposed approach significantly improved staff satisfaction, reduced stress for head nurses, and optimized healthcare resource management.

4 Bibliometric Analysis

Bibliometric analysis is a statistical evaluation of the published articles in which valuable insights are drawn from the figures and values available in published works. Although scheduling is being studied for about 3 decades now, newly developing methodologies need to be addressed with the help of bibliometric analysis. This analysis statistically analyzes nurse scheduling-related papers from the past 5 years indexed in Web of Science database. Figure 5 gives the year-wise distribution of published papers. It can be seen that after 2019, the world has felt the extreme need of scheduling strategies to be prepared for the disasters like COVID-19. Overload in the emergency departments, shortage of staff, mess up in the preplanned schedules are the major challenges which received more focus. The number of published papers related to scheduling in increasing since then.

Figure 6 tells us that Static nurse scheduling has been the most common area of research in the major area of nurse scheduling. The demand for dynamic nurse scheduling is growing rapidly due to the increasing complexity and unpredictability of modern healthcare environments. Factors such as fluctuating patient inflow, rising acuity levels, frequent emergency admissions, and sudden staff shortages have made static preplanned rosters insufficient in many scenarios. Hospitals today are expected to operate with leaner staffing models while maintaining high levels of care, which requires scheduling systems that can adapt in real-time to changes in demand and resource availability. Moreover, recent global health crises (such as COVID-19) have highlighted the limitations of traditional scheduling and accelerated the shift toward

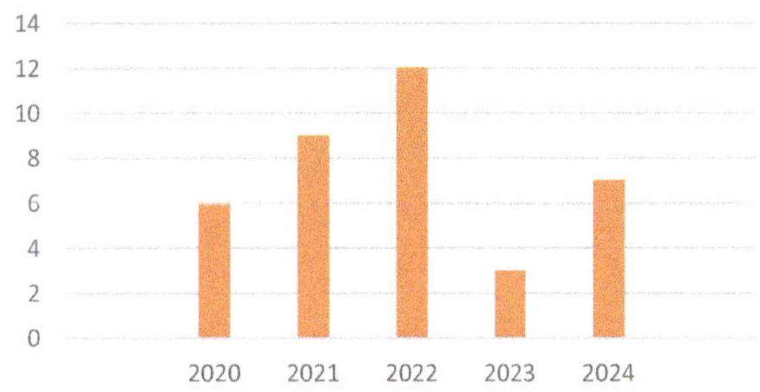

Fig. 5 Year-wise no. of papers

flexible, data-driven, and AI-assisted nurse rostering solutions. As healthcare systems continue to evolve, dynamic scheduling is no longer optional, it is becoming essential for ensuring workforce resilience and patient safety.

The methodology-wise distribution of the papers (Fig. 7) shows us that the use of hybrid approaches and integer programming has yielded good results so far. It is expected that the inventions in machine learning algorithms will elevate the scheduling strategies to a new horizon.

Figure 8 gives the distribution of papers according to the quartile ranking. Most of the papers are ranked in Q1 quartile. Figure 9 gives the distribution of papers according to the publishers. The Elsevier publication has the greatest number of published papers among all the publishers. It is notable that the difference in publications is remarkable. We also have distribution of papers according to the Journals as shown in Fig. 10. Few of the reputed journals have a good combined number of papers related to scheduling.

Between 2020 and 2024, the authors went through several research papers on nurse scheduling and looked at which optimization and programming tools were used the most. As shown in the pie chart in Fig. 11, CPLEX was the top choice, appearing in 32% of the papers, making it the go-to solver for this problem. GUROBI came next with 18%, proving to be another popular option. CLASSIC DE (Differential

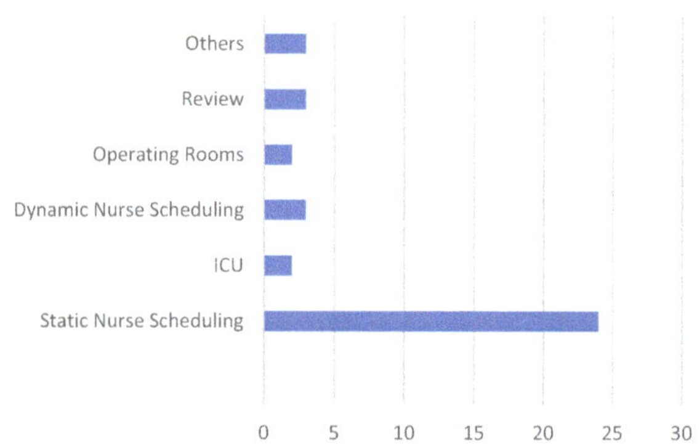

Fig. 6 Application-wise no. of papers

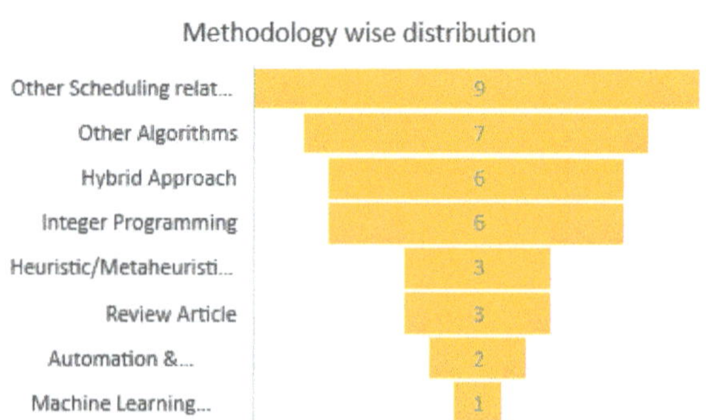

Fig. 7 Methodology-wise distribution of papers

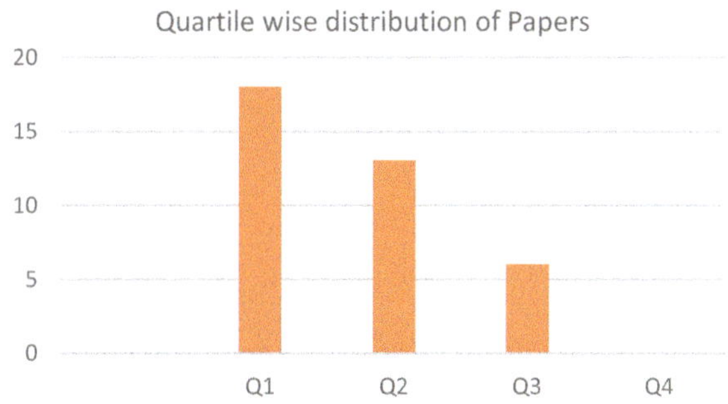

Fig. 8 Quartile-wise distribution of the papers

Evolution) and C++ were used in 14% and 7% of the studies, respectively, indicating a fair amount of interest in evolutionary algorithms and custom coding approaches. GAMS and MATLAB were each used in 11% of the papers, showing their importance in modeling and simulations. On the other hand, COIN CBC, an open-source solver, had the lowest representation at 4%, meaning that it's not as widely adopted but still has a place. Overall, this distribution shows that researchers tend to prefer commercial solvers like CPLEX and GUROBI, but alternative methods are also being explored to tackle nurse scheduling challenges.

As shown in Fig. 12, it feels like most papers focus on single-objective optimization, but their objective functions are designed to incorporate multiple objectives.

We also have country-wise distribution of research papers as shown in Figs. 13 and 14. In the field of nurse scheduling, China has the highest number of research papers (7), followed by the USA, Canada, and Iran (each with 5). Other countries like Turkey, India, and Malaysia have fewer contributions, with several countries contributing a single paper.

It would be meaningless to leave this bibliometric analysis without the distribution of citations according to each country. Figure 15 gives the representation of the countries in the World map according to the number of citations. Figure 16 gives the exact number of citations.

5 Discussion

The focus of this review was on nurse scheduling in last 5 years. All the papers available on Web of Science database were collected. Relevant articles were selected for the review. We addressed 3 research questions in this paper.

The first research question focused on the key methods and strategies used for scheduling and their relative effectiveness. After proper investigation, it was discovered that there is a vast variety of solution methodologies which were used to solve nurse scheduling problems. One notable thing which has been observed is that the hybrid approaches have been more effective to reach the optimal solutions. There

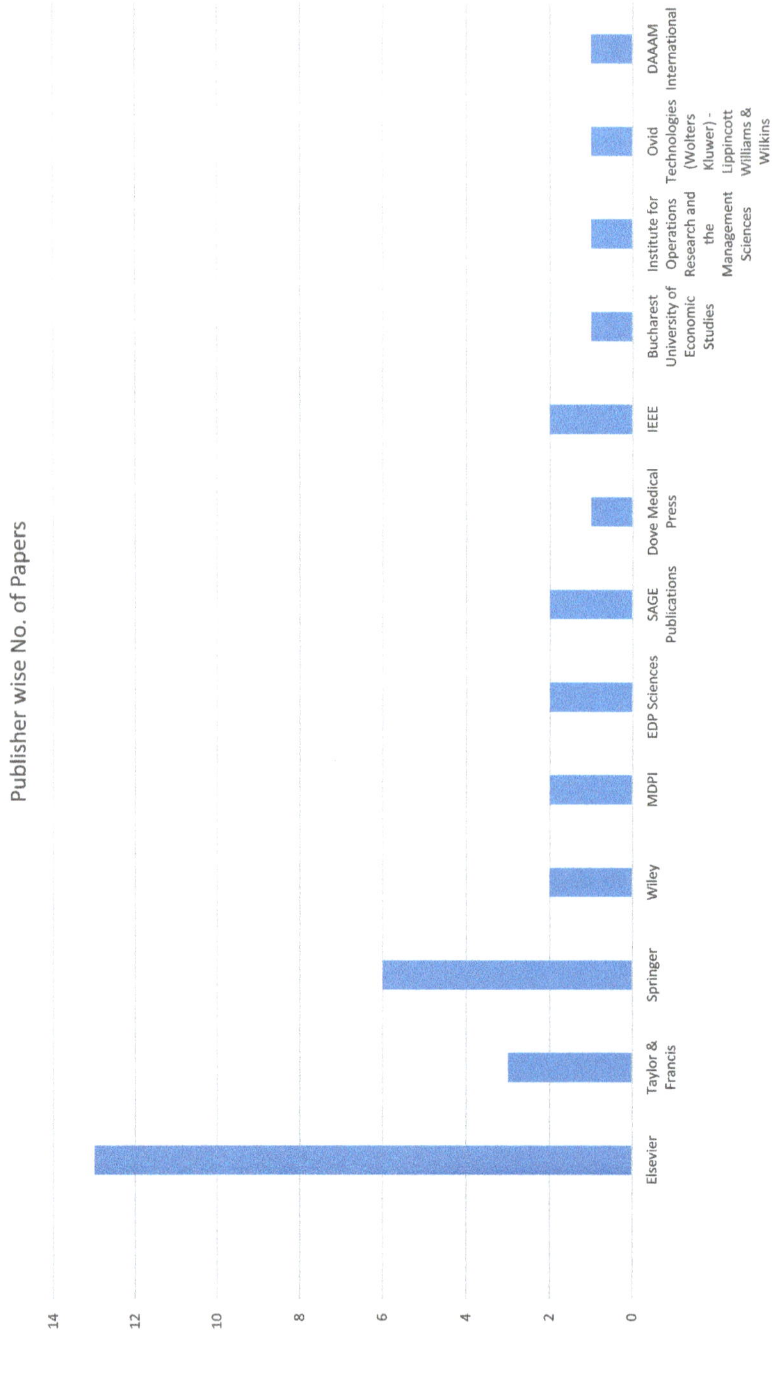

Fig. 9 Publisher-wise distribution of papers

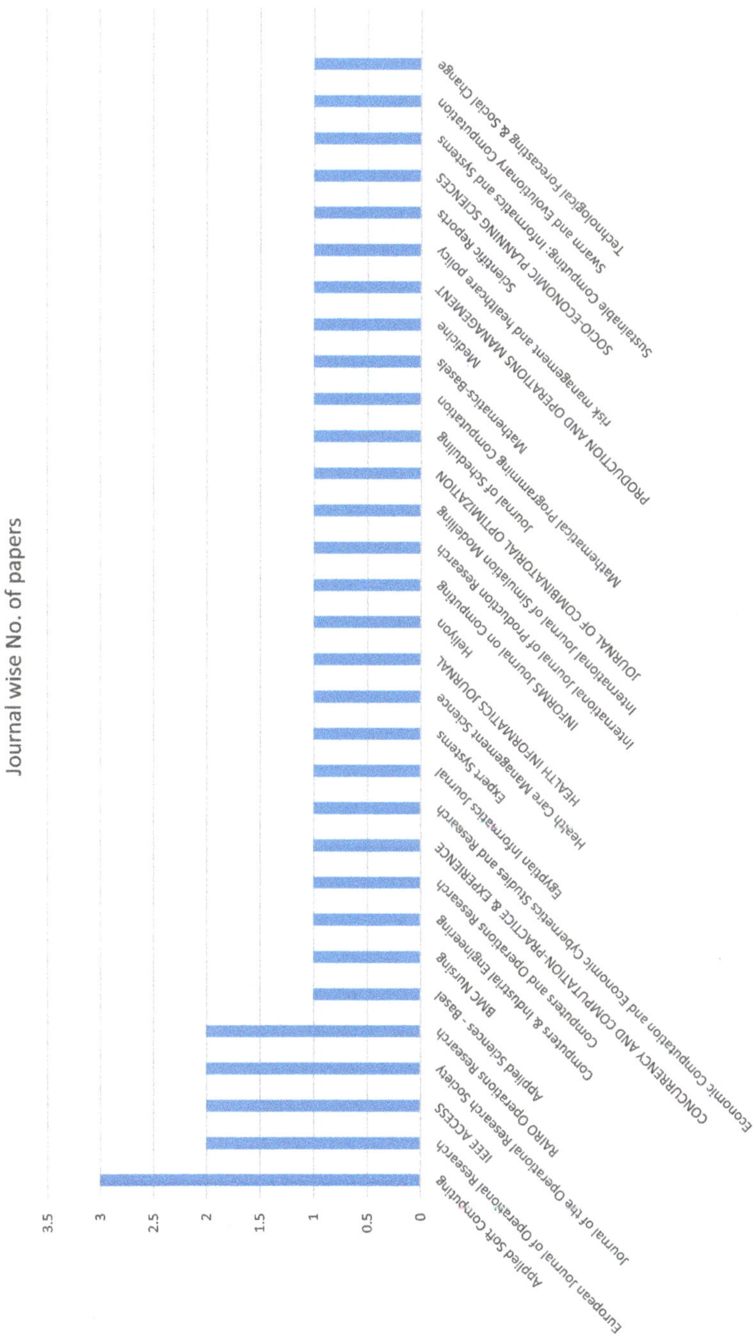

Fig. 10 Journal-wise distribution of papers

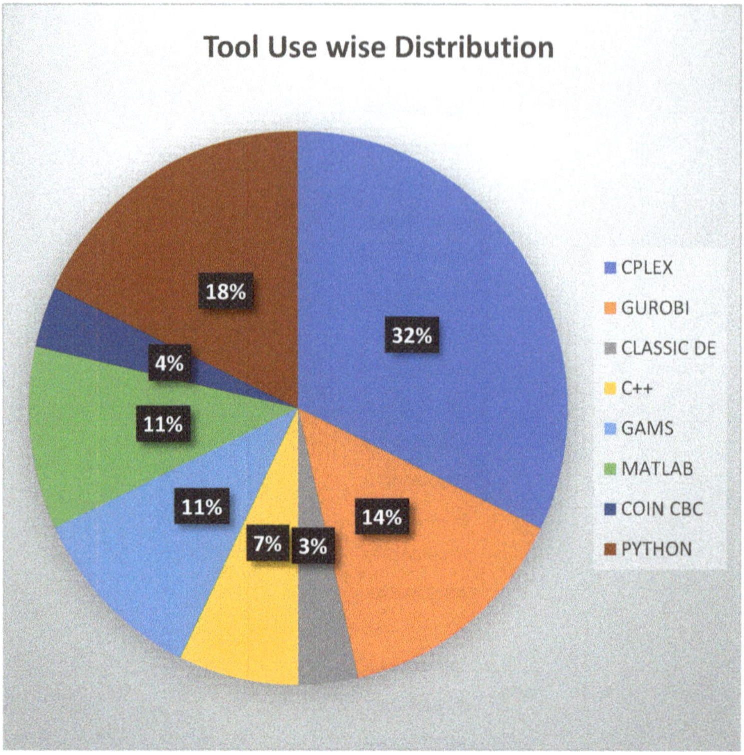

Fig. 11 Tool use-wise distribution of the papers

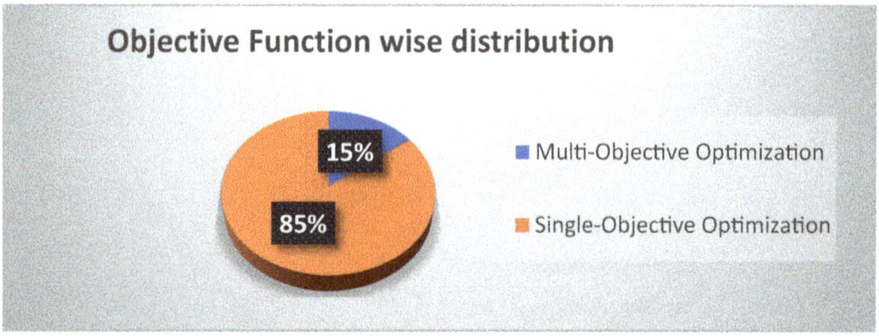

Fig. 12 Objective function-wise distribution of the papers

are other approaches to reach the optimal solution as well, but the flexibility, adaptability, scalability, and speed of hybrid approaches are the main reasons for favoring the hybrid approaches. Machine learning methods are getting developed in this field of study, and they are expected to give better results than the traditional methods such as programming methods. Figure 17 shows the timeline for the use of different algorithms in nurse scheduling problems.

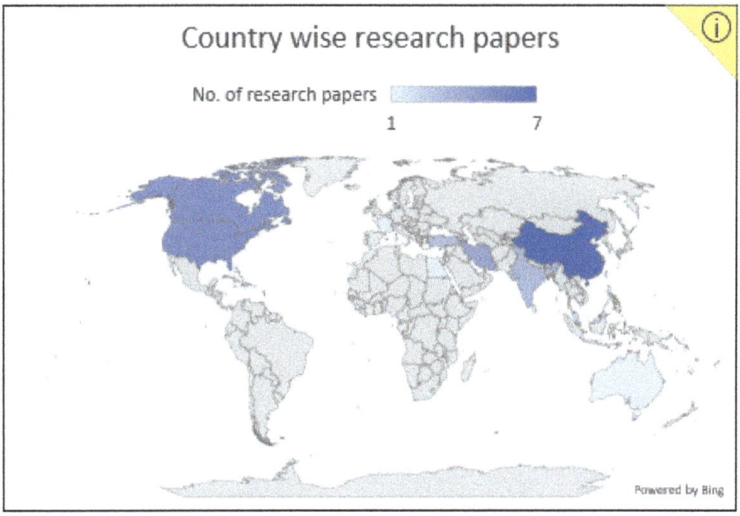

Fig. 13 Country-wise distribution of research papers in nurse scheduling

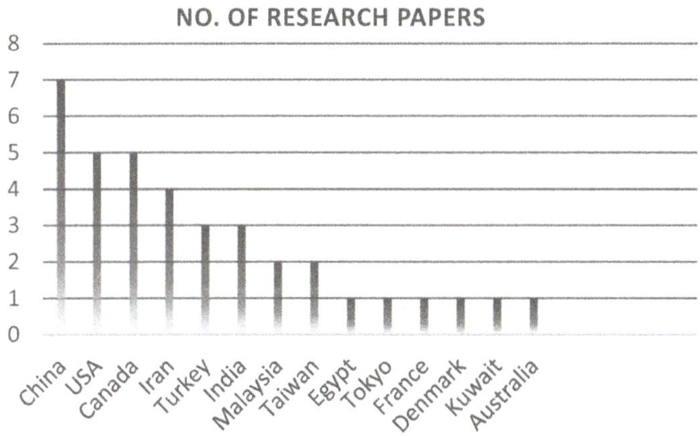

Fig. 14 Country-wise number of research papers

Second research question was based upon the handling of real-time dynamic data by the current scheduling models. Some scheduling models use dynamic staffing algorithms to incorporate situations like emergency leaves and other factors. Advanced nurse scheduling models are moving toward automation, using algorithms that can continuously optimize schedules based on changing demand. These models use real-time inputs (e.g., shift swaps, unplanned absences, or emergency procedures) and generate new schedules or shift changes with minimal human intervention.

The third research question mentioned about the effect of newly developed AI-related technology on the scheduling research and the gaps which remain for the

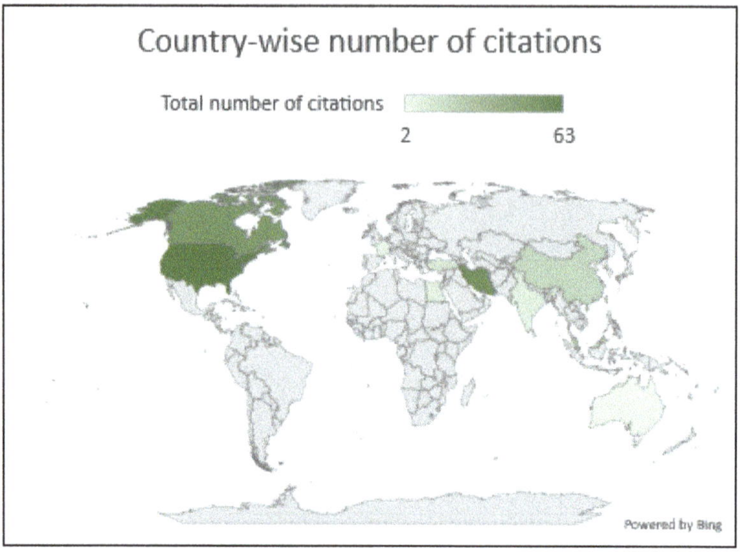

Fig. 15 Country-wise number of citations

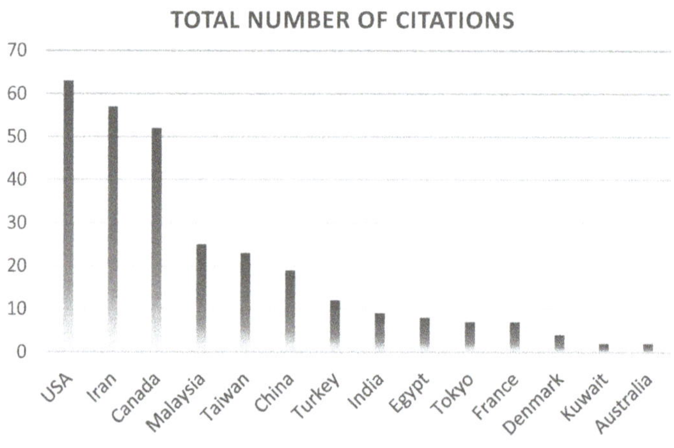

Fig. 16 Number of citations received by each country

future work. AI is being increasingly integrated into nurse scheduling to predict patient demand, forecast staffing needs, and optimize shift assignments. Machine learning models can analyze historical data, patient admission patterns, and seasonal trends to predict fluctuations in patient volumes, enabling better planning of nursing resources. Developing models that can effectively combine and analyze multi-source real-time data (patient demand, nurse availability, environmental factors) to predict staffing needs and adjust schedules on the fly remains a key challenge. Future models should incorporate human-centered constraints, such as limiting consecutive shifts,

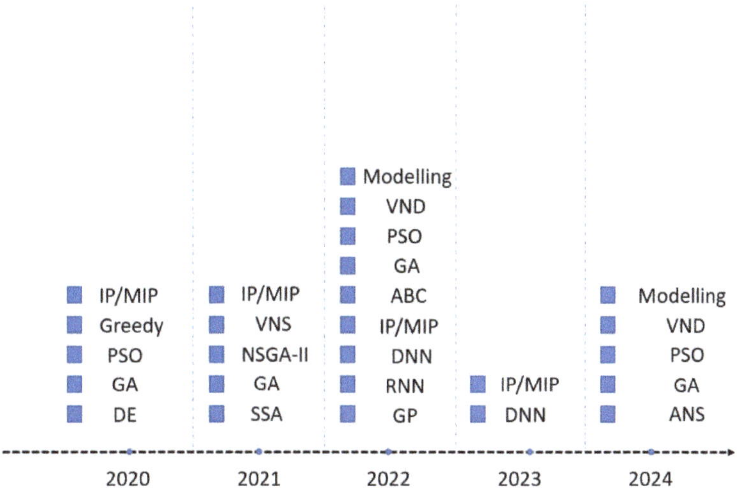

Fig. 17 Use of different algorithms

ensuring adequate rest, and optimizing workload distribution to avoid nurse burnout. AI models that can predict and adjust for these factors without negatively affecting hospital performance are crucial for the future. There is a need for research into ensuring fairness, transparency, and accountability in AI-driven nurse scheduling systems. Exploring ways to mitigate bias in AI-based scheduling and ensuring that the models do not disproportionately affect certain groups of nurses will be important.

6 Conclusion and Future Scope

The aim of this study was to study the nurse scheduling related literature in the last 5 years. Concluding remarks based on the study are as follows:

- Nurse scheduling problems in the domain of Scheduling appears to be a promising area and it has immense scope for theoretical study as well as practical implementation. AI and ML driven hybrid models are likely to become the trump cards in the scheduling. These models look promising and are getting ready to generate the best possible results.
- In recent 5 years, scheduling has gained more interest from the researchers mainly in the lockdown period where scheduling saw a lack of research in the hospitals. The growing interest is also visible from the fact that the research papers have been published in quality journals with quartile rank Q1.
- Dynamic nurse scheduling remains a critical yet under-researched area, despite its growing importance in modern healthcare systems. In today's fast-paced and unpredictable medical environments, relying solely on static scheduling is often

insufficient, as uncertainties such as sudden patient influxes, emergency cases, staff shortages, and unpredictable surgery durations demand real-time adaptability. The integration of hybrid approaches and machine learning algorithms offers significant potential for optimizing nurse schedules by balancing staff availability, patient demand, and hospital constraints.

- Hybrid models combine rule-based systems with data-driven AI techniques, while machine learning algorithms enhance scheduling by predicting staffing needs, optimizing shift assignments, and minimizing nurse fatigue.
- Given the increasing complexity of healthcare operations, advancing research in dynamic nurse scheduling is crucial for developing smarter, real-time solutions that improve hospital workflow, enhance nurse well-being, and ensure better patient outcomes.
- While recent advancements in nurse scheduling have been promising, several significant challenges continue to hinder widespread implementation and effectiveness. One of the foremost issues is scalability—real-time optimization models often struggle to perform efficiently in large, multi-department hospital systems, especially when dealing with hundreds of nurses and constantly changing demand patterns. Additionally, the growing reliance on data-driven models introduces serious concerns about data privacy and cybersecurity, as sensitive personal and health-related information must be protected under strict regulations. Another critical challenge lies in ensuring fairness and transparency within AI-based scheduling systems. If not carefully designed, these algorithms can develop unintended biases that may disproportionately affect certain groups, such as part-time staff, older nurses, or those with specific shift preferences.
- The adoption of automated scheduling systems can face cultural and operational resistance from hospital administrators and nursing staff, who may prefer manual methods or lack trust in AI-generated decisions. Addressing these concerns requires user-friendly interfaces, training, and explainable models to build confidence. From a technical perspective, balancing conflicting objectives—like minimizing cost, maximizing nurse satisfaction, and maintaining legal compliance—remains a complex optimization problem, particularly in dynamic and uncertain environments. Furthermore, adapting scheduling models to different countries' labor laws, union rules, and ethical standards poses challenges for model generalizability. Lastly, existing models often fall short in handling extreme disruptions, such as pandemics or natural disasters, which require resilient, adaptive systems capable of operating under high uncertainty and resource scarcity.
- The evolving landscape of healthcare calls for more intelligent, adaptive, and human-centered nurse scheduling systems, paving the way for several promising directions in future research. One crucial area is the integration of qualitative or "soft" factors into scheduling models—such as nurse satisfaction, emotional fatigue, teamwork dynamics, and psychological well-being. These elements, although difficult to quantify, significantly influence job performance and retention, and their incorporation can be enabled through sentiment analysis, survey-based feedback, or psychometric tools.

- Another promising direction is the development of interdisciplinary scheduling frameworks that go beyond nurses to include physicians, technicians, and support staff, promoting better coordination in multidisciplinary care units like ICUs or emergency departments. With increasing reliance on automation, explainable AI (XAI) is gaining attention to ensure that the decision-making processes behind AI-generated schedules are transparent and interpretable to users. This could enhance trust and accountability in clinical environments where human oversight is critical.
- Additionally, the use of wearable technologies and IoMT (Internet of Medical Things) holds the potential to create adaptive scheduling systems that respond to real-time biometric data—such as fatigue levels, stress indicators, or sleep patterns—enabling dynamic, personalized shift recommendations that improve both nurse health and patient safety.
- Cross-hospital or regional nurse-sharing networks represent another future avenue, where intelligent systems could coordinate staffing across a network of hospitals during emergencies or pandemics, optimizing system-wide resource utilization.
- The inclusion of sustainability metrics is also a timely direction—future models could be designed to minimize energy use (e.g., night shifts) and support green hospital initiatives. Moreover, hybrid models that combine AI techniques with traditional optimization and simulation-based methods are likely to offer robust solutions capable of handling multiple conflicting objectives.

As healthcare systems continue to embrace digital transformation, these future directions can help create smarter, fairer, and more resilient nurse scheduling systems that align with both operational goals and human values.

References

1. Schoenfelder, J., Bretthauer, K.M., Wright, P.D., Coe, E.: Nurse scheduling with quick-response methods: Improving hospital performance, nurse workload, and patient experience. Eur. J. Oper. Res. **283**, 390–403 (2020). https://doi.org/10.1016/j.ejor.2019.10.047
2. Pahlevanzadeh, M.J., Jolai, F., Goodarzian, F., Ghasemi, P.: A new two-stage nurse scheduling approach based on occupational justice considering assurance attendance in works shifts by using Z-number method: A real case study. RAIRO-Oper. Res. **55**, 3317–3338 (2021). https://doi.org/10.1051/ro/2021157
3. A sensing-based patient classification framework for efficient patient-nurse scheduling - ScienceDirect, https://www.sciencedirect.com/science/article/pii/S2210537923000100, last accessed 2025/02/03
4. Multi-agent deep Q-network-based metaheuristic algorithm for Nurse Rostering Problem - ScienceDirect, https://www.sciencedirect.com/science/article/pii/S2210650224000853, last accessed 2025/02/03
5. A column generation-based algorithm for midterm nurse scheduling with specialized constraints, preference considerations, and overtime - ScienceDirect, https://www.sciencedirect.com/science/article/pii/S0305054821003191, last accessed 2025/02/03

6. A Survey of the Nurse Rostering Solution Methodologies: The State-of-the-Art and Emerging Trends | IEEE Journals & Magazine | IEEE Xplore, https://ieeexplore.ieee.org/document/9780256, last accessed 2025/02/03
7. A Review of the Scheduling Problem within Canadian Healthcare Centres, https://www.mdpi.com/2076-3417/12/21/11146, last accessed 2025/02/03
8. Zhang, X., Huang, D.-S., Guan, P.: Nursing Scheduling Mode and Experience from the Medical Teams in Aiding Hubei Province During the COVID-19 Outbreak: A Systematic Scoping Review of 17 Studies. Risk Manag. Healthc. Policy. **14**, 1805–1813 (2021). https://doi.org/10.2147/RMHP.S302156
9. Full article: New results on integrated nurse staffing and scheduling: The medium-term context for intensive care units, https://www.tandfonline.com/doi/full/10.1080/01605682.2020.1806742, last accessed 2025/02/03
10. Pérez, E., Dzubay, D.P.: A scheduling-based methodology for improving patient perceptions of quality of care in intensive care units. Health Care Manag. Sci. **24**, 203–215 (2021). https://doi.org/10.1007/s10729-021-09544-6
11. Böðvarsdóttir, E.B., Bagger, N.-C.F., Høffner, L.E., Stidsen, T.J.R.: A flexible mixed integer programming-based system for real-world nurse rostering. J. Sched. **25**, 59–88 (2022). https://doi.org/10.1007/s10951-021-00705-7
12. Full article: Short-Term nurse schedule adjustments under dynamic patient demand, https://www.tandfonline.com/doi/full/10.1080/01605682.2022.2039566, last accessed 2025/02/03
13. A Dedicated Pricing Algorithm to Solve a Large Family of Nurse Scheduling Problems with Branch-and-Price | INFORMS Journal on Computing, https://pubsonline.informs.org/doi/10.1287/ijoc.2023.0019, last accessed 2025/02/03
14. Legrain, A., Omer, J., Rosat, S.: A rotation-based branch-and-price approach for the nurse scheduling problem. Math. Prog. Comp. **12**, 417–450 (2020). https://doi.org/10.1007/s12532-019-00172-4
15. Turhan, A.M., Bilgen, B.: A mat-heuristic based solution approach for an extended nurse rostering problem with skills and units. Socio Econ. Plan. Sci. **82**, 101300 (2022). https://doi.org/10.1016/j.seps.2022.101300
16. A New Mathematical Model for Staff Scheduling in the Operating Room Based on their Preferences, https://www.researchgate.net/publication/366384956_A_New_Mathematical_Model_for_Staff_Scheduling_in_the_Operating_Room_Based_on_their_Preferences, last accessed 2025/02/03
17. Hassani, M.R., Behnamian, J.: A scenario-based robust optimization with a pessimistic approach for nurse rostering problem. J. Comb. Optim. **41**, 143–169 (2021). https://doi.org/10.1007/s10878-020-00667-0
18. Digitalisation for optimising nursing staff demand modelling and scheduling in nursing homes, https://ideas.repec.org/a/eee/tefoso/v164y2021ics004016252031338x.html, last accessed 2025/02/03
19. Sea lion attacking-based deer hunting optimization algorithm for dynamic nurse scheduling in health care sector contribution of hybrid algorithm in cloud - Sarkar - 2022 - Concurrency and Computation: Practice and Experience - Wiley Online Library, https://onlinelibrary.wiley.com/doi/full/10.1002/cpe.7249, last accessed 2025/02/03
20. Developing two heuristic algorithms with metaheuristic algorithms to improve solutions of optimization problems with soft and hard constraints: An application to nurse rostering problems - ScienceDirect, https://www.sciencedirect.com/science/article/pii/S1568494620302763, last accessed 2025/02/03
21. HSSAGA: Designation and scheduling of nurses for taking care of COVID-19 patients using novel method of Hybrid Salp Swarm Algorithm and Genetic Algorithm - ScienceDirect, https://www.sciencedirect.com/science/article/pii/S1568494621003720, last accessed 2025/02/03
22. Artificial Bee Colony Algorithm with Nelder-Mead Method to Solve Nurse Scheduling Problem, https://www.mdpi.com/2227-7390/10/15/2576, last accessed 2025/02/03

23. Chen, Z., Dou, Y., De Causmaecker, P.: Neural networked-assisted method for the nurse rostering problem. Comput. Ind. Eng. **171**, 108430 (2022). https://doi.org/10.1016/j.cie.2022.108430
24. Chen, Z., De Causmaecker, P., Dou, Y.: A Combined Mixed Integer Programming and Deep Neural Network-Assisted Heuristics Algorithm for the Nurse Rostering Problem, https://papers.ssrn.com/abstract=4020057, (2022). doi:10.2139/ssrn.4020057
25. Harb, H., Abboud, A., Kwekha Rashid, A.S., Saad, G., Abouaissa, A., Idoughmar, L., AlAkkoumi, M.: An intelligent optimization strategy for nurse-patient scheduling in the Internet of Medical Things applications. Egyptian Informatics Journal. **25**, 100451 (2024). https://doi.org/10.1016/j.eij.2024.100451
26. Nagayoshi, M., Tamaki, H.: A Dynamic Nurse Scheduling Approach Using Reinforcement Learning to Address Sudden Absences of an Unknown Nurse
27. Muklason, A., Kusuma, S.D.R., Riksakomara, E., Premananda, I.G.A., Anggraeni, W., Mahananto, F., Tyasnurita, R.: Solving Nurse Rostering Optimization Problem using Reinforcement Learning - Simulated Annealing with Reheating Hyper-heuristics Algorithm. Procedia Comput. Science. **234**, 486–493 (2024). https://doi.org/10.1016/j.procs.2024.03.031
28. Abdelghany, M., Yahia, Z., Eltawil, A.B.: A new two-stage variable neighborhood search algorithm for the nurse rostering problem. RAIRO-Oper. Res. **55**, 673–687 (2021). https://doi.org/10.1051/ro/2021027
29. A 2-Stage Approach for the Nurse Rostering Problem | IEEE Journals & Magazine|IEEE Xplore, https://ieeexplore.ieee.org/document/9805588, last accessed 2025/02/03
30. Chen, P.-S., Huang, W.-T., Chen, G.Y.-H., Dang, J.-F., Yeh, E.-C.: Constructing modified variable neighborhood search approaches to solve a nurse scheduling problem. Int. J. Prod. Res. **62**, 7186–7204 (2024). https://doi.org/10.1080/00207543.2024.2320707
31. An online stochastic algorithm for a dynamic nurse scheduling problem - ScienceDirect, https://www.sciencedirect.com/science/article/pii/S0377221718308038, last accessed 2025/02/03
32. D'Souza, B., Rao, S.S., Muthana, C.G., Bhageerathy, R., Apuri, N., Chandrasekaran, V., Prabhavathi, D., Renukaradhya, S.: ROTA: A system for automated scheduling of nursing duties in a tertiary teaching hospital in south India. Health Informatics J. **27**, 14604582211001426 (2021). https://doi.org/10.1177/14604582211001426
33. Valdano, E., Poletto, C., Boëlle, P.-Y., Colizza, V.: Reorganization of nurse scheduling reduces the risk of healthcare associated infections. Sci. Rep. **11**, 7393 (2021). https://doi.org/10.1038/s41598-021-86637-w
34. Deriving the minimum staff number requirement for intelligent staff scheduling: An efficient constructive method and application - Cao - 2022 - Expert Systems - Wiley Online Library, https://onlinelibrary.wiley.com/doi/full/10.1111/exsy.12975, last accessed 2025/02/03
35. Kirli Akin, H., Ordu, M.: A Novel Simulation-Based Two-Stage Optimization Approach for Nurse Planning. Int. j. simul. model. **21**, 591–602 (2022). https://doi.org/10.2507/IJSIMM21-4-618
36. Patient-controlled use of nonphysician providers: Appointment scheduling in mixed-provider settings - Taiwo - 2023 - Production and Operations Management - Wiley Online Library, https://onlinelibrary.wiley.com/doi/full/10.1111/poms.14000, last accessed 2025/02/03
37. An examination of the usefulness of a quantitative appraisal method in nursing human resource management in primary hospital operating rooms: An example of integrated collaborative scheduling - PubMed, https://pubmed.ncbi.nlm.nih.gov/38728512/, last accessed 2025/02/03
38. Home health care nurse routing and scheduling problem considering ergonomic risk factors - ScienceDirect, https://www.sciencedirect.com/science/article/pii/S2405844023111042, last accessed 2025/02/03
39. Booker, L.A., Mills, J., Bish, M., Spong, J., Deacon-Crouch, M., Skinner, T.C.: Nurse rostering: understanding the current shift work scheduling processes, benefits, limitations, and potential fatigue risks. BMC Nurs. **23**, 295 (2024). https://doi.org/10.1186/s12912-024-01949-2
40. El Adoly, A.A., Gheith, M., Nashat Fors, M.: A new formulation and solution for the nurse scheduling problem: A case study in Egypt. Alex. Eng. J. **57**, 2289–2298 (2018). https://doi.org/10.1016/j.aej.2017.09.007

41. M'Hallah, R., Alkhabbaz, A.: Scheduling of nurses: A case study of a Kuwaiti health care unit. Oper. Res. Health Care. **2**, 1–19 (2013). https://doi.org/10.1016/j.orhc.2013.03.003
42. Yasmine, A., Yassine, O., Farouk, Y., Hicham, C.: Workload balancing for the nurse scheduling problem: A real-world case study from a French hospital. Socio Econ. Plan. Sci. **95**, 102046 (2024). https://doi.org/10.1016/j.seps.2024.102046

A Review of Bilevel Optimization Methods, Emerging Applications, and Recent Advancements

Dhaval Pujara and Ankur Sinha

Abstract This chapter presents a comprehensive review of techniques proposed in the literature for solving bilevel optimization problems encountered in various real-life applications. Bilevel optimization is an appropriate choice for hierarchical decision-making situations, where a decision-maker needs to consider a possible response from stakeholder(s) for each of its actions to achieve his own goals. Mathematically, it leads to a nested optimization structure, in which a primary (leader's) optimization problem contains a secondary (follower's) optimization problem as a constraint. Various forms of bilevel problems, including linear, mixed-integer, single-objective, and multi-objective, are covered. For bilevel problem-solving methods, various classical and evolutionary approaches are explained. Along with an overview of various areas of applications, two recent considerations of bilevel approach are introduced. The first application involves a bilevel decomposition approach for solving general optimization problems, and the second application involves neural architecture search (NAS), which is a prime example of a bilevel optimization problem in the area of machine learning.

Keywords Bilevel optimization · Bilevel optimization-based decomposition · Neural architecture search

1 Introduction

Bilevel optimization is a special class of optimization problems characterized by a unique structure, where the primary optimization problem contains an additional optimization problem, i.e., a secondary optimization problem, as one of its constraints. In literature [1, 2], the primary and secondary optimization problems are

D. Pujara (✉) · A. Sinha
Indian Institute of Management Ahmedabad (IIM Ahmedabad), Ahmedabad, India
e-mail: dhavalp@iima.ac.in

A. Sinha
e-mail: asinha@iima.ac.in

S. Mukhopadhyay et al. (eds.), *Contemporary Advancement in Evolutionary Multi-objective Optimization*, Genetic and Evolutionary Computation,
https://doi.org/10.1007/978-981-95-7908-2_6

referred to as the upper level and lower level optimization problems, respectively. From a game theory point of view, this setup represents a hierarchical decision-making scenario, where two entities, leader (upper level) and follower (lower level), are associated with each other in a way that the leader needs to think of every possible response from the follower for each of its strategies for achieving own goals. Thus, the leader's action depends on the reaction from the follower. This mechanism is shown in Fig. 1, representing the inter-linkage between upper and lower levels. Leader has multiple choices or strategies from the upper level decision space $\mathcal{X}$. The decision vector associated with any strategy is represented as an upper level decision vector x^p ($p = 1, ..., |\mathcal{X}|$). Leader, aware of the set of possible responses from the follower, considers different decisions (x^1, x^2,...) from $\mathcal{X}$. Follower responds to each decision of leader (x^1, x^2,...) with an appropriate decision (y^1, y^2,...) from the lower level decision space $\mathcal{Y}$. For each upper level decision x^p, follower performs a lower level parametric optimization using a suitable optimization method, treating x^p as a parameter. This process efficiently explores the lower level decision space $\mathcal{Y}$ and identifies the best (optimal) lower level response y^p to x^p. A pair (x^p, y^p) represents a feasible solution to the upper level optimization problem provided that it satisfies all the constraints in the problem. The overall aim is to find the best solution for upper level optimization problem, a pair (x^*, y^*) that minimizes/maximizes the upper level objective function. Thus, in bilevel optimization problem, the leader (upper level) and the follower (lower level) have their own objectives and constraints and both aim to find optimal solutions while functioning in a described decision-making scenario. Here, we discussed a situation where leader usually has complete knowledge of follower's strategies, while the follower only observes the leader's decision and then reacts optimally. Interestingly, there can be various kinds of uncertainties in bilevel problems, for instance, parameter uncertainty, variable uncertainty and decision-making uncertainty. Some of the studies in this direction are [3–10].

Bilevel optimization has retained the interest of researchers and practitioners as many real-world situations can be formulated as bilevel optimization problems. For

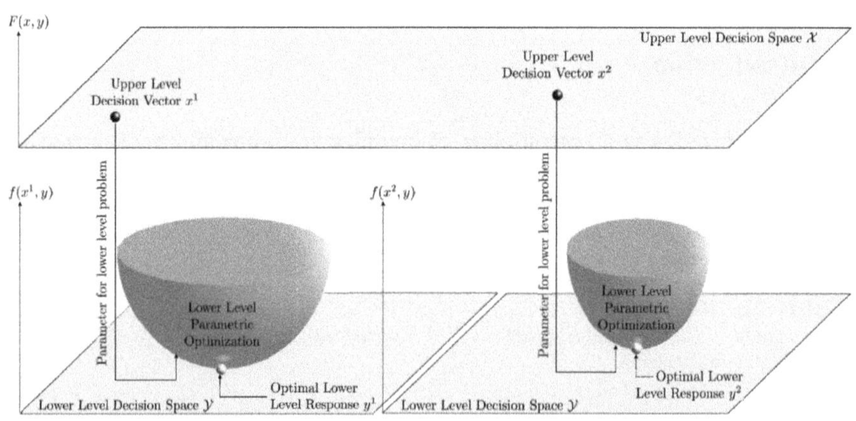

Fig. 1 Sketch of decision-making mechanism in bilevel optimization problem

instance, pricing models, network design for supply chain management, and competitors operating in the same market are several examples that naturally fit into the bilevel framework as their problem structure involves hierarchical decision-making. Bilevel approach has also been adopted in policy making, where the central authority aims to achieve Social, Technological, Economic, Environmental, or Political (STEEP) enhancement by controlling the actions of consumers through suitable policy norms (strategies). For example, farmers often overuse fertilizers to increase crop yields, which negatively impact the environment through land and water pollution. In [11–13], authors propose a bilevel model-based policy that encourages farmers to reduce fertilizer usage and indirectly prevents pollution. According to policy norms, government provides an incentive to farmers if they use fertilizers within a specific range. With this incentive, farmers get the same net profit without overusing fertilizers. Thus, bilevel concept is applied in public sector to make positive impact on environment without compromising the interests of farmers. Apart from the policy formation aimed at regulating STEEP factors, bilevel optimization has been widely applied to homeland security problems such as interdiction of nuclear weapons [14, 15], border security [16–18], defending terror attacks [19–21], and protecting critical infrastructure [22–24]. In recent times, computer scientists are using bilevel programming for tuning the hyperparameters of various algorithms. Hyperparameters are configuration variables that determine the structural and learning characteristics of an algorithm, e.g., in neural networks, hyperparameters include the learning rate, number of layers, number of neurons per layer, and batch size; in K-means clustering, the number of clusters; in decision trees, the maximum depth and minimum number of samples per leaf; and in evolutionary algorithms, the population size, crossover and mutation probabilities. The values of hyperparameters are set before starting the model training process. Although hyperparameter values are typically set by the user, several studies [25–29] have developed bilevel-based methods to perform the same task.

From a problem-solving perspective, the hierarchical decision-making structure often leads to non-convex and disconnected feasible regions, which imparts the NP-hard property to bilevel problems and make them difficult to solve mathematically [30, 31]. A study by Deng [32] provides a proof that no polynomial-time algorithm exists for solving linear bilevel optimization problems. Due to these challenges, traditional mathematical programming-based optimization methods fail to solve complex bilevel optimization problems efficiently. Apart from that, metaheuristic algorithms (e.g., genetic algorithm, simulated annealing, etc.) are found to be effective in handling some of the inherent difficulties in bilevel problems. Therefore, a combined application of classical methods and metaheuristic algorithms has provided promising results for certain classes of challenging bilevel optimization problems [33–37]. Overall, literature suggests various approaches and methods for solving bilevel problems and at the same time, due to the difficulties exhibited by bilevel problem structure, there is significant scope for developing new methods to address these problems more efficiently.

A network map in Fig. 2 depicts the applied and theoretical research topics addressed using bilevel optimization and interconnections among them. Each link

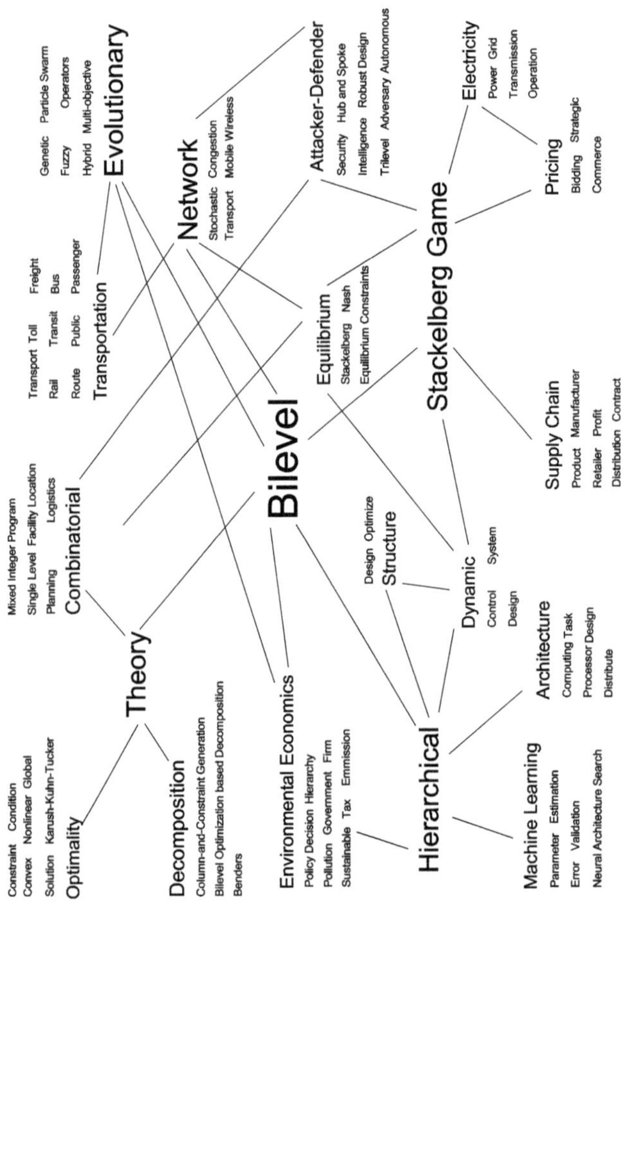

Fig. 2 Network map depicting connections between research topics addressed using bilevel optimization

in map connects a subtopic to a higher level topic represented by a larger font size. Along with bilevel optimization methods, this chapter provides a review of recent advancements in bilevel optimization research and its application to real-world problems. Accordingly, we start with a basic definition and mathematical presentations of bilevel optimization problem along with the most frequently used terminologies in bilevel optimization. Later, the classical and evolutionary approaches for solving bilevel optimization problems are explained. The method sections are followed by mixed-integer bilevel optimization and multiobjective bilevel optimization sections. Next section provides a summary of real-life problems that are addressed using bilevel optimization. In the same section, two recent applications of bilevel optimization approach in (i) solving single level optimization problems using a bilevel optimization-based decomposition method, and (ii) neural architecture search are explained in detail. The chapter concludes with a discussion on future research directions in this field.

2 Bilevel Formulation

In this section, we formally define the bilevel optimization problem using a mathematical formulation. The notation and setup for bilevel optimization are as follows: the upper level (leader's) optimization problem contains a lower level (follower's) optimization problem nested within it as a constraint. The upper level decision vector is denoted by x ($x = (x_1, ..., x_n) \in \mathbb{R}^n$) and the lower level decision vector is represented by y ($y = (y_1, ..., y_m) \in \mathbb{R}^m$). There are separate sets of objective functions and constraints for the upper level ($F(x, y) : \mathbb{R}^n \times \mathbb{R}^m \to \mathbb{R}$ and $G_p(x, y) : \mathbb{R}^n \times \mathbb{R}^m \to \mathbb{R},\ p = 1, ..., P$) and the lower level ($f(x, y) : \mathbb{R}^n \times \mathbb{R}^m \to \mathbb{R}$ and $g_q(x, y) : \mathbb{R}^n \times \mathbb{R}^m \to \mathbb{R},\ q = 1, ..., Q$) optimization problems. The lower level problem is a parametric optimization problem to be solved optimally for lower level decision variables with upper level decision variables passed as parameters. A lower level solution is considered valid if it satisfies all the upper level constraints, and then the complete solution (x, y) acts as a feasible solution to the upper level problem. With this background, a mathematical formulation of the bilevel problem is provided in Definition 1 as follows:

Definition 1 A bilevel optimization problem with upper and lower level optimization tasks can be formulated as

$$\text{“}\min_{x,y}\text{”} \quad F(x, y) \tag{1}$$

$s.t.$

$$y \in \operatorname*{argmin}_{y}\{f(x, y) : g_q(x, y) \le 0,\ q = 1, \ldots, Q\} \tag{2}$$

$$G_p(x, y) \le 0,\ p = 1, \ldots, P \tag{3}$$

Above bilevel formulation may also include equality constraints, which are exempted here for brevity. Also, the upper level and lower level decision variables may be integer-valued. However, unless explicitly stated otherwise, we assume these variables to be real and continuous throughout our discussion.

The lower level problem (Constraint 2) can be written using a set-valued mapping, which is unknown a priori as follows:

Definition 2 Let $\Psi : \mathbb{R}^n \rightrightarrows \mathbb{R}^m$ be a set-valued mapping,

$$\Psi(x) = \underset{y}{\operatorname{argmin}}\{f(x, y) : g_q(x, y) \leq 0, \ q = 1, \ldots, Q\} \tag{4}$$

In a nutshell, $\Psi(x)$ represents a mapping that returns the values of lower level variables $y = (y_1, \ldots, y_m)$ for a given x. With $\Psi(x)$, the bilevel optimization problem (Definition 1) can be expressed as a general constrained optimization problem as follows:

$$\underset{x,y}{\text{“min”}} \quad F(x, y) \tag{5}$$

$$s.t.$$

$$y \in \Psi(x) \tag{6}$$

$$G_p(x, y) \leq 0, \ p = 1, \ldots, P \tag{7}$$

In Definitions 1 and 2, quotes have been used in upper level objective function to reflect the ambiguity that arises in decision-making at upper level when multiple lower level optimal solutions exist for any given upper level decision vector. In this scenario, the decision-maker at upper level faces uncertainty regarding which optimal solution will be picked by lower level decision-maker. A solution selected from the multiple lower level optimal solutions may or may not benefit the leader, depending on whether the follower's behavior is cooperative or adversarial. The problem becomes fully defined when it is clear which kind of solution will be selected by the lower level in such situations. This aspect is taken into consideration by defining the optimistic and pessimistic positions as follows:

2.1 Optimistic Position

In optimistic position, from the set of multiple lower level optimal solutions, the follower is expected to select a solution that leads to the best objective function value for the upper level or leader's optimization problem. This reflects a certain degree of cooperation between leader and follower. The follower's choice function under the optimistic assumption, $\Psi^o(x)$, can be defined as follows:

$$\Psi^o(x) = \underset{y}{\operatorname{argmin}}\{F(x, y) : y \in \Psi(x)\} \tag{8}$$

Accordingly, the formulation of bilevel optimization problem with optimistic choice function (8) is provided below:

$$\min_{x,y} \quad F(x, y) \tag{9}$$

$$s.t.$$

$$y = \Psi^{o}(x) \tag{10}$$

$$G_p(x, y) \leq 0, \quad p = 1, \dots, P \tag{11}$$

A bilevel formulation with optimistic position (9–11) is guaranteed to have an optimal solution when it satisfies several mathematical properties mentioned in the theorem below:

Theorem 1 *If the objective function and constraints of bilevel optimization problem (F, f, G_p, g_q) are sufficiently smooth, the constraint region Φ is non-empty and compact, and the Mangasarian-Fromowitz constraint qualification holds at all points, then the problem is guaranteed to have an optimistic bilevel optimum, provided there exists a feasible solution.*

For detailed discussions about the existence of optimistic bilevel optimum and additional results on optimality conditions, readers may refer to [2, 38–44].

2.2 *Pessimistic Position*

In pessimistic position, from the set of multiple lower level optimal solutions, the follower is expected to select a solution that leads to the least favorable outcome for the upper level optimization problem (i.e., leader receives the minimum benefit from the selected lower level optimal solution compared to other lower level optimal solutions). This shows a lack of cooperation between leader and follower. Under pessimistic setting, the choice function of the follower, $\Psi^{p}(x)$, can be defined as follows:

$$\Psi^{p}(x) = \underset{y}{\operatorname{argmax}}\{F(x, y) : y \in \Psi(x)\} \tag{12}$$

Accordingly, the formulation of bilevel optimization problem with pessimistic choice function (12) is provided below:

$$\min_{x,y} \quad F(x, y) \tag{13}$$

$$s.t.$$

$$y = \Psi^{p}(x) \tag{14}$$

$$G_p(x, y) \leq 0, \quad p = 1, \dots, P \tag{15}$$

Between optimistic and pessimistic positions, optimistic position is relatively more tractable as it is possible to reduce the optimistic bilevel formulation to a single-level optimization problem using the variational inequality corresponding to lower level problem, provided the lower level problem is convex. In the case of pessimistic position, such a straightforward single-level conversion is not possible. As a result, pessimistic bilevel optimization requires explicitly tracking of the lower level optimal solution that yields the worst outcome for the upper level problem, which effectively makes the bilevel problem a three-level task. There are certain studies that approximate the pessimistic problems through perturbed optimistic problems, for instance [45]. Moreover, the pessimistic bilevel problem is guaranteed to have an optimal solution under stronger assumptions provided in the theorem below:

Theorem 2 *If the objective function and constraints of bilevel optimization problem* (F, f, G_p, g_q) *are sufficiently smooth, the constraint region* Φ *is non-empty and compact, and the set-valued mapping* $\Psi^p(x)$ *is lower semi-continuous for all upper level solutions, then the problem is guaranteed to have a pessimistic bilevel optimum.*

For details on existence of pessimistic bilevel optimum and additional results on optimality conditions, readers may refer to [2, 44, 46–51]. For a comprehensive overview of the evolution and advancements in bilevel optimization, readers can refer to review papers [1, 52–61] and books/edited volumes [2, 62–69]. Bilevel optimization is closely related to mathematical program with complementarity constraints (MPCC). For discussion of differences between the two classes of problems, refer to [70].

A summary of the terms and notations commonly used in the bilevel optimization literature is provided in Table 1 .

3 Bilevel Problem-Solving Methods

In literature, bilevel optimization problems are typically addressed using the classical or evolutionary approaches. Classical approach includes single-level reduction methods, duality methods, descent methods, penalty function methods, and trust region methods. Broadly, these methods are referred to as mathematical programming-based techniques. Single-level reduction methods transform the bilevel problem into single-level optimization problem by replacing the lower level problem with its Karush-Kuhn-Tucker (KKT) conditions, a set-valued mapping, or the lower level optimal value function. Since bilevel problems belong to the class of complex optimization problems, mathematical programming-based techniques are generally applied to bilevel problems that are mathematically well-behaved, typically those with linear, quadratic, or convex objective functions and constraints. Apart from that, strong assumptions such as continuous differentiability and lower semi-continuity are also very common for classical methods. For complex bilevel problems that do not comply with the strong assumptions and favorable mathematical properties (e.g.,

Table 1 Key terms and notations used in bilevel optimization literature

Terms	Notation (s)	Description
Decision vectors	$x \in X$ $y \in Y$	Upper level decision vector (x) and decision space (X). Lower level decision vector (y) and decision space (Y).
Objectives	F f	Upper level objective function(s). Lower level objective function(s)
Constraints	$G_p,\ p = 1, \ldots, P$ $g_q,\ q = 1, \ldots, Q$	Upper level constraint functions. Lower level constraint functions.
Lower level feasible region	$\Omega : X \rightrightarrows Y$	$\Omega(x) = \{y : g_q(x, y) \leq 0\ \forall\ q\}$, represents the lower level feasible region for any given x.
Constraint region (relaxed feasible set)	$\Phi = \text{gph}\ \Omega$	$\Phi = \{(x, y) : G_p(x, y) \leq 0\ \forall\ p,\ g_q(x, y) \leq 0\ \forall\ q\}$, represents the region satisfying both upper and lower level constraints.
Lower level reaction set	$\Psi : X \rightrightarrows Y$	$\Psi(x) = \{y : y \in \text{argmin}_{y \in Y}\{f(x, y) : y \in \Omega(x)\}\}$, shows the lower level optimal solution(s) for a given x.
Inducible region (feasible set)	$I = \text{gph}\ \Psi$	$I = \{(x, y) : (x, y) \in \Phi,\ y \in \Psi(x)\}$, represents the set of upper level decision vectors and corresponding lower level optimal solution(s) belonging to feasible constraint region.
Choice function	$\psi : X \to Y$	$\psi(x)$ represents the solution chosen by the follower for any x. It becomes important in case of multiple lower level optimal solutions.
Optimal value function	$\varphi : X \to \mathbb{R}$	$\varphi(x) = \min_y\{f(x, y) : y \in \Omega(x)\}$, represents the minimum lower level function value corresponding to a given x.

convexity, continuity, differentiability, etc.), evolutionary approaches are employed, which include nature-inspired metaheuristic algorithms such as genetic algorithm, particle swarm optimization, etc. In this section, we briefly explain various bilevel problem-solving methods covered in the classical and evolutionary approaches.

3.1 Karush-Kuhn-Tucker Conditions-Based Single-Level Reduction

In the scenario of lower level problem being convex and sufficiently regular, bilevel optimization problem can be reformulated as a single-level optimization problem by replacing the lower level problem with its Karush-Kuhn-Tucker (KKT) conditions. Accordingly, the problem in Definition 1 can be reduced to the formulation given by (16)–(21).

$$\min_{x,\, y,\, \lambda} \quad F(x, y) \tag{16}$$

$$s.t.$$

$$G_p(x, y) \leq 0, \;\; p = 1, \ldots, P \tag{17}$$

$$\nabla_y L(x, y, \lambda) = 0 \tag{18}$$

$$g_q(x, y) \leq 0, \;\; q = 1, \ldots, Q \tag{19}$$

$$\lambda_q \, g_q(x, y) = 0, \;\; q = 1, \ldots, Q \tag{20}$$

$$\lambda_q \geq 0, \;\; q = 1, \ldots, Q \tag{21}$$

where

$$L(x, y, \lambda) = f(x, y) + \sum_{q=1}^{Q} \lambda_q \, g_q(u, l).$$

The KKT conditions-based formulation [(16)–(21)] sometimes becomes difficult to handle as Lagrangian constraint (18) may induce non-convexity, even when the bilevel problem follows convexity and regularity conditions. Also, the complementary conditions (20), inherently being combinatorial, make the reduced single-level problem a mixed-integer program. In the case of linear bilevel problem, the Lagrangian constraint remains linear and complementary conditions are linearized using combinatorial variables, which provides the reduced formulation in Mixed-Integer Linear Program (MILP) form. Sometimes, the complementary conditions are also handled using Special Ordered Sets (SOS). To solve the MILP, vertex enumeration [71–73] and Branch-and-Bound (B&B) [74, 75] approaches are considered in literature. Though B&B methods become slow as the number of integer variables increases, these methods are successfully applied for solving single-level reductions of bilevel problems having linear-quadratic [76] and quadratic-quadratic [77,

78] structures as well. An extended KKT approach has been proposed in [79] for linear bilevel problems. For approximate KKT conditions in the context of bilevel programs, readers may refer to [80].

3.2 Duality-Based Single-Level Reduction

If the lower level problem is convex and satisfies strong duality, such as through Slater's condition, the bilevel optimization problem can be equivalently reformulated as a single-level program by replacing the lower level problem with its dual. This technique is broadly applicable to various classes of bilevel problems that meet the convexity and strong duality criteria.

To illustrate this approach, we will use a general linear bilevel program with continuous variables as an example. However, the underlying principle extends naturally to other convex bilevel formulations where strong duality holds.

$$\min_{x,y} \; c_x^\top x + c_y^\top y \tag{22}$$

$$s.t.$$

$$A_x x + A_y y \geq a \tag{23}$$

$$y \in \underset{y}{\operatorname{argmin}} \left\{ d^\top y : B_x x + B_y y \geq b \right\} \tag{24}$$

Next, let us write an alternative single-level reformulation of the bilevel program in (22–24) using the duality-based approach. The dual of the lower level linear program is given as follows:

$$\min_{\lambda} \; (b - B_x x)^\top \lambda \quad s.t. \quad B_y^\top \lambda = d, \lambda \geq 0.$$

Now the linear bilevel program in (22–24), can be reformulated as a single-level problem as follows:

$$\min_{x,y,\lambda} \; c_x^\top x + c_y^\top y \tag{25}$$

$$s.t.$$

$$A_x x + A_y y \geq a \tag{26}$$

$$B_x x + B_y y \geq b \tag{27}$$

$$B_y^\top \lambda = d, \lambda \geq 0 \tag{28}$$

$$d^\top y \leq (b - B_x x)^\top \lambda. \tag{29}$$

Note that the formulation contains a bilinear constraint, which may not be easy to linearize as the complementary slackness conditions because the bilinear terms do not equate to zero. Refer to [81] for an overview of using such approaches for practical problem-solving.

3.3 Value Function-Based Single-Level Reduction

This approach uses φ-mapping (optimal value function in Table 1) to obtain the optimal objective function value for lower level problem $f(x, y)$. This replaces the lower level optimization problem (Constraint 2 in Definition 1) with a φ-mapping constraint and leads to a single-level optimization problem formulation (30)–(33) as follows:

$$\min_{x,y} \quad F(x, y) \tag{30}$$

$$s.t.$$

$$f(x, y) \leq \varphi(x) \tag{31}$$

$$g_q(x, y) \leq 0, \;\; q = 1, \ldots, Q \tag{32}$$

$$G_p(x, y) \leq 0, \;\; p = 1, \ldots, P \tag{33}$$

In practice, the optimal value function $\varphi(x)$ is not known a priori. Therefore, one cannot readily solve the above problem (30)–(33) to arrive at bilevel optimum. Instead, φ-mapping-based algorithms estimate $\varphi(x)$ during the iterations of the algorithm. Some approaches in this direction are [37, 82–85].

3.4 Descent Methods

In bilevel optimization setup, the aim of descent approach is to identify the descent direction that leads to decrease in the upper level objective function value such that a new point remains feasible. For this feasible point, apart from maintaining the feasibility for upper level problem, it should also be ensuring optimality for the lower level problem. Finding such descent direction can be challenging. To address this, various approaches such as approximating the gradient of upper level objective [86] and formulating auxiliary problems [87, 88] are suggested in the literature.

In [87], assuming a unique optimal solution, linear independence, convexity, and second-order sufficiency for the lower level problem, the authors propose solving an auxiliary linear-quadratic bilevel program to determine the steepest descent direction. For the lower level problem,

$$\min_{x,y} \quad f(x, y)$$

$$s.t.$$

$$g_i(x, y) \leq 0, i = 1, \ldots, I$$

Let the Lagrangian be represented as follows, where $\mathcal{I}(x) \subseteq \{1, \ldots, I\}$ represents the indices corresponding to the binding constraints at the lower level optimum.

$$L(x, y, \lambda) = f(x, y) + \sum_{i \in \mathcal{I}(x)} \lambda_i g_i(x, y)$$

The authors propose solving a linear-quadratic bilevel program to compute the steepest descent direction $z \in \mathbb{R}^n$ at a point x, given $y \in \Psi(x)$ and a uniquely determined multiplier vector λ ensured by the assumptions.

$$\begin{aligned}
&\min_{z,w} \nabla_x F(x, y)^\top z + \nabla_y F(x, y)^\top w \\
&s.t. \\
&w \in \operatorname*{argmin}_{w} \Big\{ \left(z^\top, w^\top\right) \nabla^2_{(x,y)} L(x, y, \lambda)(z, w) \\
&\qquad\qquad s.t. \\
&\qquad\qquad\quad \nabla_y g_i(x, y) w \leq -\nabla_x g_i(x, y) z, \quad i \in \mathcal{I}(x) \\
&\qquad\qquad\quad \nabla_y f(x, y) w = -\nabla_x f(x, y) z + \nabla_x L(x, y, \lambda) z \Big\} \\
&-1 \leq z \leq 1
\end{aligned}$$

In the above formulation, the upper level objective function denotes the directional derivative of $F(x, y)$ along (z, w), which is minimized to obtain the steepest descent direction. The quadratic program yields w, indicating how the lower level solution shifts as the upper level variable x moves along z.

3.5 Penalty Function Approach

Methods from the class of penalty function approach address the bilevel optimization problem by solving a series of unconstrained optimization problems. The unconstrained problem is formed by incorporating a penalty term that measures the violation of the constraints. The penalty term takes the value zero for a feasible solution (i.e., eliminates the penalty term) and takes a positive value (in minimization case) for infeasible solutions (i.e., penalizes the objective function). Penalty function approach was initially implemented in [89, 90]. Both studies replace the lower level problem with a penalized problem; however, the resultant structure maintains the hierarchy of bilevel optimization, which still remains difficult to solve. Later, [91] proposed a double penalty method. As the name suggests, this method penalizes both upper and lower level objective functions using the penalty approach. Then, a penalized lower level problem is replaced with corresponding KKT conditions to reduce the bilevel problem to a single-level problem, which is subsequently solved using a penalization technique. There are several studies where lower level optimization problem is directly replaced with corresponding KKT conditions and then penalization approach is applied for solving the single-level problem. In [92, 93], penalty function approach is considered to solve linear bilevel optimization problems. Former study [92] converts a bilevel problem to a penalized bilevel problem and solves a series of bilevel problems to achieve the bilevel optimum. Later study [93] performs a single-level reduction using the lower level KKT conditions and updates the upper level objective function by adding complementary slackness conditions with a penalty; then, a reduced single-level problem is solved using series of linear programs.

3.6 Trust Region Approach

Trust region methods perform the local approximation of objective function in the neighborhood of current solution, known as the *trust region*, where the approximation is assumed to be reliable. These methods are iterative in nature, i.e., build and solve the local approximation of model function in step-by-step manner to reach at the bilevel optimum. These methods are found effective for handling nonlinearity, non-convexity, or non-regularities in bilevel problems. Trust region method was firstly considered in [94] to solve nonlinear bilevel optimization problem with a lower level problem having convex objective function and linear constraints. The study does not contain constraints at the upper level. Later, [95] proposed a more general approach of performing the local approximation of bilevel problem with a linear program at the upper level and a linear variational inequality at the lower level; and then, the solution procedure involves trust region and line search mechanisms to reach the bilevel optimum over the iterations. Other study [96] suggests to approximate the bilevel problem with a linear-quadratic bilevel problem and then solve its reduced single-level formulation as a mixed-integer program.

3.7 Evolutionary Approach

Evolutionary approach is preferred to deal with complexities such as nonlinearity, non-convexity, discontinuity, and non-differentiability in bilevel problems. This approach uses metaheuristic algorithm(s), nature-inspired or intelligence-based effective strategies for sampling the solution space, to solve the bilevel problem. Various metaheuristic algorithms used in the literature for solving bilevel optimization problems are genetic algorithm [97–105], particle swarm optimization [106–108], differential evolution [109, 110], scatter search algorithm [111], etc. Recent comprehensive reviews of metaheuristic approaches for bilevel optimization can be found in [112, 113].

Based on the problem-solving approach, the bilevel evolutionary methods are classified into three categories: (i) nested methods, (ii) single-level reduction, and (iii) meta-modeling. All three types of methods are briefly discussed in this section.

Nested methods solve bilevel problems in nested form (1–3), where lower level problem is solved for each sampled upper level solution. There are two ways considered for solving bilevel problems using nested methods: (i) upper level problem is sampled using metaheuristic algorithm and the corresponding lower level problem is solved using classical method; it is known as hybrid-nested approach and (ii) metaheuristic algorithms are applied at both upper and lower levels to obtain the complete solution. The decision on selecting any approach depends on the complexity of lower level optimization problem. For example, bilevel problems with regular lower level optimization problem are addressed using the hybrid-nested approach [97–100], and the approach of applying metaheuristic algorithm at both levels is considered in the case of bilevel problem with complex lower level optimization problem [35, 102, 106, 111, 114–116].

The purpose and mechanism of single-level reduction methods, in the context of evolutionary approach, are similar to what we discussed in Sect. 3.1, where bilevel problem is reduced to a single-level problem by replacing the lower level problem with its KKT conditions, provided that the lower level problem satisfies certain regularity conditions. Most of the time, it is observed that reduced single-level problems continue to sustain various complexities, because of which solving the reduced single-level problem is also not a straightforward task. In such a scenario, evolutionary approach is useful due to better capability of metaheuristic algorithms in handling non-regularities. For example, one of the earliest studies based on evolutionary approach, [101], applies the single-level reduction over linear bilevel problem and solves the single-level problem using genetic algorithm, where chromosomes emulate the vertex points. Another study [117] uses a simplex-based genetic algorithm to solve single-level formulation corresponding to linear-quadratic bilevel problem. Other approaches that use evolutionary algorithms with KKT-based reduction and often rely on additional optimization principles are [80, 103–105, 108].

Meta-modeling approach is used in optimization when actual function evaluation is very time-consuming or computationally expensive. A meta-model or surrogate model is an approximation of the original model and it is relatively fast to evaluate. Solving a bilevel optimization problem, nested formulation (1–3), leads to a large number of evaluations as we solve lower level problem for each upper level solution, which is a computationally expensive task. If the lower level optimization problem is complex, then this task becomes even more expensive. In this scenario, a combined application of meta-model and metaheuristic algorithm has been observed as an efficient strategy. Reaction set mapping and optimal lower level function value, Ψ-mapping and φ-mapping provided in Table 1, are often approximated with meta-models. For any given upper level solution vector x, Ψ-mapping provides the lower level decision vector and φ-mapping returns the optimal value of lower level optimization problem. In general, none of the mappings are available at the beginning of bilevel problem solving. Therefore, initially, lower level problem is solved for a few upper level solutions, and later the required mapping is approximated ($\hat{\Psi}(x)$ or $\hat{\varphi}(x)$) using the lower level problem elements (y or f) and the corresponding upper level solutions. For complex lower level optimization problems, it is hard to approximate the entire mapping, hence, practitioners consider the iterative meta-modeling approach, where required mapping is approximated locally several times over the iterations. After approximating the lower level problem with $\hat{\Psi}(x)$ or $\hat{\varphi}(x)$ mapping, the reduced single-level problem is solved using metaheuristic algorithm. For evolutionary approach where meta models are used, readers can refer to [36, 37, 84, 109, 118–120].

4 Mixed-Integer Bilevel Optimization

The formulation of mixed-integer bilevel optimization problem includes the constraints of a general bilevel problem, as given in (1–3) in Definition 1 or (5–7) in Definition 2, along with additional constraints that require one or more variables to take integer values only (i.e., any $x_i \in \mathbb{Z}$ and/or $y_j \in \mathbb{Z}$, where x_i is a component of x and y_j is a component of y). One of the earlier studies [121] on mixed-integer bilevel optimization examined the properties of mixed-integer linear bilevel programs and conditions for the existence of optimal solutions for the problem configuration. In classical optimization literature, branch-and-bound is one of the most commonly used techniques for handling integer variables in MILP. However, branch-and-bound cannot be directly applied to Mixed-Integer Bilevel Linear Problems (MIBLP) due to several challenges in fathoming, as discussed in [122], which also proposed a branch-and-bound technique with strict fathoming conditions and several heuristics to effectively handle problems with more number of integer variables. Later, in a subsequent study, [123] authors developed an implicit enumeration scheme for mixed-integer linear bilevel problems under the condition that integer variables can take binary values only. For the continuous-discrete bilevel problems (where $x \in \mathbb{R}$ and $y \in \mathbb{Z}$), [124] proposed a cutting plane method that applies the *Chvatal-Gomory* cut. Studies, such as [125–127], use benders-decomposition-based techniques to solve bilevel problems with mixed-integer upper level variables and continuous linear lower level variables, while [83] uses value function reformulation for mixed integers at both levels. Fischetti et al. [128] proposed a general-purpose algorithm for MIBLP, where both upper and lower levels have linear constraints and objective functions, and some/all variables take integer values. In a subsequent study [129], the authors used intersection cuts to solve MIBLP. In [130], a column and constraint generation-based decomposition algorithm is proposed for the single-level formulation corresponding to MIBLP. For branch-and-cut implementations that capitalize on useful cuts, readers can refer to [131, 132]. Generalized benders decomposition and an extended KKT transformation are discussed in [133] for a mixed-integer nonlinear multilevel model corresponding to the zonal pricing problem in electricity markets. In addition to classical methods, evolutionary approaches have also been used to solve mixed-integer bilevel problems in [65, 134–140]. For the latest developments in discrete bilevel optimization, readers can refer to [15, 133, 141–145].

5 Multiobjective Bilevel Optimization

In practical scenario, the leader and/or follower may have multiple objectives. This leads to the general formulation of multiobjective bilevel optimization problem below.

Definition 3 For the upper level objective function $F(x, y) : \mathbb{R}^n \times \mathbb{R}^m \rightarrow \mathbb{R}^s$ and lower level objective function $f(x, y) : \mathbb{R}^n \times \mathbb{R}^m \rightarrow \mathbb{R}^t$, the multiobjective bilevel problem is formulated in (34–36) as follows:

$$\min_{x,y} \quad F(x, y) = (F_1(x, y), ..., F_s(x, y)) \tag{34}$$

$$s.t.$$

$$y \in \underset{y}{\text{argmin}}\{f(x, y) = (f_1(x, y), ..., f_t(x, y)) :$$

$$g_q(x, y) \leq 0, \ q = 1, \ldots, Q\} \tag{35}$$

$$G_p(x, y) \leq 0, \ p = 1, \ldots, P \tag{36}$$

In above formulation, $G_p : \mathbb{R}^n \times \mathbb{R}^m \rightarrow \mathbb{R}$, $p = 1, \ldots, P$ represent the upper level constraints, and $g_q : \mathbb{R}^n \times \mathbb{R}^m \rightarrow \mathbb{R}$, $q = 1, \ldots, Q$ denote the lower level constraints. Additionally, there may be integer restrictions on upper and/or lower level variables.

In literature, multiobjective bilevel optimization has received very scant treatment compared to single-objective bilevel optimization, primarily due to computational and decision-making complexities associated with this class of problems. The optimality conditions for multiobjective bilevel programs are discussed in [146–148]. In [149], an $\epsilon-$constraint technique is applied at both levels of multiobjective bilevel problem, which results into the $\epsilon-$constraint bilevel problem. The $\epsilon-$parameter is passed by the decision-maker, and the problem is solved by replacing the lower level problem with its KKT conditions. Eichfelder [150, 151] addressed the multiobjective bilevel optimization problems using a classical approach, where the author used a numerical optimization technique and an adaptive exhaustive search method to solve lower level and upper level problems, respectively. The use of exhaustive search method makes this procedure time-consuming and less effective for the large-scale problems. A number of studies involve multiple objectives at the lower level and a single objective at the upper level, and are referred to as semi-vectorial bilevel optimization problems. A common approach to handle such problems is to scalarize the lower level using parameters λ, and then consider these parameters as a part of upper level decision vector while converting the lower level to a single objective optimization problem [152]. This reduces the overall semi-vectorial problem into a single-objective bilevel optimization problem (i.e., one objective at both levels). This idea has also been exploited for bilevel problems with multiple objectives at both levels [153]. Other papers on semi-vectorial bilevel optimization are [152, 154–158].

With the popularity of evolutionary algorithms in solving multiobjective (single-level) optimization problems during 1990s and 2000s, in late 2000s, researchers started applying the metaheuristic algorithms for solving multiobjective bilevel problems. Yin [99] formulated the transportation planning and management problem as multiobjective bilevel program, where upper level has multiobjective and lower level contains a single objective, and solved it using a nested genetic algorithm. Later, Halter and Mostaghim [159] proposed a particle swarm optimization-based nested strategy to solve the multiobjective bilevel program representing the chemical system. To deal with several complexities such as nonlinearity, non-convexity, and non-differentiability in multiobjective bilevel problems, Sinha et al. [4, 160–162] and Deb & Sinha [33, 163, 164] have suggested evolutionary and hybrid approaches to handle multiple objectives at both the upper and the lower levels. An approximation approach of the set-valued mapping has been performed in the context of multiobjective problems as well [165]. Many of the above studies in the context of multiple objectives at both levels take an optimistic position from the decision-making point of view, i.e., among the Pareto-optimal solutions from the lower level, the upper level decision-maker freely chooses the one(s) that are most suitable at the upper level. In many real-world scenarios this may not be an actual scenario. For studies that address such concerns with discussions on optimistic/pessimistic Pareto-optimal frontiers and decision-making uncertainties, readers may refer to [4, 50, 55, 166, 167].

In [33], authors have also developed a suite of test problems to evaluate and compare the performance of various algorithms designed for solving multiobjective bilevel problems. The latest survey on the main approaches for multiobjective bilevel optimization is available in [168]. Readers may also refer to [50, 169–175] for other developments in multiobjective bilevel optimization.

6 Real-Life Applications of Bilevel Optimization

Problems arising in various domains such as economics, supply chain management, engineering, and management, among others, often exhibit structures that are well-suited to be addressed using the bilevel optimization approaches. In this section, we briefly describe real-life problems addressed using the bilevel optimization and provide a list of relevant reference studies.

(1) Toll Setting Problem: This problem belongs to the class of network problems. In this problem, the authority, who acts as a leader, wants to optimize the toll rates for the network of roads by considering the behavior of network users, the followers. The studies relevant to this problem are [176–187].
(2) Environmental Economics: In this class of problems, the authority wants to tax an organization or individuals who consume a particular commodity to generate the revenue. Excessive consumption of this commodity leads to adverse environmental impacts. Therefore, knowing the consumption behavior at various tax rates, authority aims to control the use of commodity by deciding a tax rate that

leads to the prevention of environmental pollution without letting consumers lose too much revenue. Variants of this problem are covered in [12, 188–191].

(3) Interdiction Problems: Interdiction problems are a class of optimization problems under attack or disruption, where an interdictor (attacker) tries to disrupt or degrade the performance of a system, while a defender (operator) tries to optimize system performance despite the disruption. Interdiction is often considered for the nodes or arcs of a network. Depending on who is the leader and who is the follower, interdiction problems are often formulated as attacker-defender or defender-attacker problems. These problems may also lead to multilevel optimization (beyond two levels) when network design and protection are taken into account. Some studies in this direction are [192–199].

(4) Facility or Hub Location: When deciding on the location of a facility or hub, firms may consider the potential responses of its competitors or customers. This scenario represents the Stackelberg game decision-making environment, which can be effectively be addressed using bilevel optimization approach. Location problems under the risk of attacks are commonly addressed as interdiction problems as stated above. In [200], study considers a scenario where a firm enters a market by locating a new facility, and competitor responds to that by adjusting the attractiveness of its current facilities. Other facility problems addressed using bilevel optimization can be retrieved from [125, 135, 194, 201–212].

(5) Chemical Industry: For the chemical process, practitioners often want to decide the state variables and quantity of reactants to achieve the optimal output. In this setup, optimizing the output is the upper level problem and the lower level problem appears as an entropy function minimization problem. This problem is considered in [213–215].

(6) Optimal Design: The bilevel approach is frequently employed in topology optimization and structural engineering, where the objective is to determine the optimal shape, material distribution, and quantity of material that minimize the overall weight or cost, or equivalently, maximize the structural strength and stiffness at the upper level. The upper level problem typically encapsulates design decisions subject to physical and performance constraints, such as limits on displacements, stresses, or contact forces, ensuring structural integrity and feasibility. At the lower level, the problem commonly manifests as a potential energy minimization, compliance maximization, worst-case disturbance identification, or as a variational inequality representing the equilibrium state of the structure under applied loads and boundary conditions. This hierarchical formulation elegantly couples design optimization with mechanical equilibrium, allowing the structural behavior to be consistently reflected in design updates. For this class of problems, readers may refer to [216–222].

(7) Inverse Optimal Control: This problem is mainly observed in computer vision, remote sensing, robotics, and related fields. In control theory, one of the major tasks is to obtain the performance index or reward function that fits best on a given dataset. This task is associated with inverse optimal control theory, where

one obtains the calculation of the cause based on the given result. Such requirement requires solving a parameter estimation problem with an optimal control problem. This bilevel nature problem is studied in [223–226].

(8) Principal-Agent Problems: It is a classical problem observed in economics area, where a principal (leader) subcontracts a job to an agent (follower). In this problem, it is important for principal to take the agent's preferences into account while designing the incentive scheme as the agent is expected to act in his own interests rather than those of the principal. This setup matches with the bilevel decision-making mechanism and can be correlated with real-life situation experienced with doctor-patient, employer-employee, politician-voters, corporate board-shareholders, etc. The studies related to this problem are [227–230].

(9) Energy Networks and Market: After liberalization of the electricity sector and the introduction of energy markets, private power generation companies, market operators, and transmission system operators have become part of the decision-making process in energy sector. The nature of interaction and sequence in which these entities make decisions match with a Stackelberg-type environment, i.e., bilevel optimization scenario. Therefore, in recent times, bilevel optimization has been extensively employed to address problems associated with energy networks and markets [67]. The vulnerability of power systems and the security of power grids under disruptive threats are discussed in [231, 232]. Problems of power generation, transmission, and capacity planning and expansion are studied in [233–235]. A recent survey on application of bilevel optimization in electricity market is available in [236]. Apart from electricity market, energy management in the gas market using bilevel optimization is discussed in [237–239].

Next, we delve into the recent applications of bilevel approach in more detail.

6.1 Bilevel Optimization-Based Decomposition

Bilevel optimization-based decomposition (BOBD) is a recently developed decomposition method, covered in [240, 241], for effectively solving the complex and large-scale single-level or general optimization problems. As the name suggests, BOBD method uses a decomposition strategy to transform a general optimization problem into an equivalent bilevel optimization problem. This decomposition allows the use of effective bilevel optimization techniques in solving general optimization problems. The motivation and procedure of formulating a general optimization problem as a bilevel optimization problem and solving it using bilevel optimization methods are discussed in this section. A mathematical formulation of general optimization problem is provided below.

Definition 4 The single-level optimization problem, with decision variables $x = (x_1, \ldots, x_n)$, objective function $F(x)$, and constraints set $G(x)$, can be formulated as follows:

$$\min_{x} \quad F(x) \tag{37}$$

$$\text{s.t.} \quad G_r(x) \leq 0, \quad r = 1, \ldots, R \tag{38}$$

The single-level optimization problem (37–38) corresponding to the real-world optimization system mostly contains a large number of decision variables and constraints, which leads to large-scale optimization problem. Additionally, the objective function $F(x)$ and constraints $G(x)$ may contain several terms that lead to non-regularities such as nonlinearity, non-convexity, non-differentiability, etc. Solving such complex optimization problems, characterized by large n and R along with non-regularities, using a classical approach or evolutionary approach alone does not yield effective solutions generally. That is because, classical methods can handle the large-scale scenario only when problems follow certain regularity conditions related to linearity, convexity, differentiability, etc. On the other side, metaheuristic algorithms can effectively handle the non-regularities but their performance deteriorates as the size of the problem increases. Thus, both classical and evolutionary approaches, when employed independently, fail to handle the large-scale and non-regular scenarios simultaneously. However, most of the time, real-world optimization problems hold both of these complexities simultaneously. Thus, in practical scenario, both approaches become ineffective since only one approach can be employed to solve single-level problem at a time. However, BOBD method considers hybrid approach that allows applying both classical and metaheuristic algorithms simultaneously to solve bilevel optimization problem corresponding to the single-level optimization problem. The working mechanism of BOBD method is discussed next.

Any constrained single-level optimization problem (Definition 4) has mainly two components: objective function (37) and constraints (38). Both of these components are basically the mathematical functions of decision variables x, i.e., $F(x)$ and $G(x)$. In other words, $F(x)$ and $G(x)$ are mathematical expressions composed of terms containing decision variables (constants are omitted in this discussion). Hence, decision variables are the basic component of optimization problem, and that is why they are the source of complexities, if any exist. To understand this fact, consider the numerical example provided below.

$$\begin{aligned}
\min_{x} \quad & x_1 - 2x_2^3 + 4x_3 \\
\text{s.t.} \quad & -x_1 - x_3 \leq -5; \\
& x_2^2 + x_3 \leq 4; \\
& 0 \leq x_1,\ x_2,\ x_3 \leq 10
\end{aligned}$$

In the above numerical example, term $-2x_2^3$ makes the objective function non-convex and term x_2^2 causes nonlinearity in the second constraint. Imagining the above example without the terms containing x_2 decision variable would turn the problem into a linear optimization problem since all the remaining terms with decision variables x_1 and x_2 are linear. Thus, variable x_2 is the source of complexity in the instance. Based on this analysis, each variable (x_1, x_2, x_3) can be tagged as complexity-causing variable $\{x_2\}$ or complexity-soothing variable $\{x_1, x_3\}$. In this way, current exercise, *complexity analysis*, shows the role of variables in the complexity of problem.

The decomposition strategy of BOBD method is built on the complexity analysis procedure discussed above using an example. It performs complexity analysis for each decision variable x_i ($x_i \in x$, $i \in [1, n]$) and classifies x_i into upper level variables category u ($u \subseteq x$) if x_i is complexity-causing variable; otherwise, when x_i is complexity-soothing variable, x_i is classified into lower level variables category l ($l \subseteq x$). Thus, $u \cup l = x$, with $u \cap l = \emptyset$. This classification further allows us to write objective function $F(x)$ and constraints $G_r(x)$ in terms of upper and lower level variables u and l, i.e., $F(u, l)$ and $G_r(u, l)$. Later, based on the presence of upper and lower level variables, each constraint $G_r(u, l)$ is separated as the upper level constraint $G_{r^+}(u, l)$ or the lower level constraint $G_{r^-}(u, l)$. Using this variable and constraint classifications, u/l, $G_{r^+}(u, l)/G_{r^-}(u, l)$, $F(u, l)$, the single-level optimization problem (37–38) can be represented in the form of bilevel optimization problem as follows:

$$\min_{u,l} \quad F(u, l) \tag{39}$$

$$s.t. \quad l \in \underset{l}{\operatorname{argmin}}\{F(u, l) : G_{r^-}(u, l) \leq 0\} \tag{40}$$

$$G_{r^+}(u, l) \leq 0 \tag{41}$$

In above formulation, the objective functions at both upper and lower levels are same, i.e., $F(u, l)$. Hence, solving this bilevel formulation (39–41) would lead to the same optimal solution as the formulation (37–38).

For the numerical example considered in ongoing section, an equivalent bilevel formulation can be obtained using a decomposition strategy of BOBD method as follows. Among the three decision variables $x_1, x_2, x_3 \in x$, variable x_2, complexity-causing variable, is classified into the upper level variables category (i.e., $x_2 = u_1 \in u$); and variables x_1 and x_3, complexity-soothing variables, are classified into the lower level variables category (i.e., $x_1, x_3 = l_1, l_2 \in l$). This classification leads to $x = (x_1, x_2, x_3) = (l_1, u_1, l_2)$. Using this classification, the numerical instance, in the form of single-level optimization problem, can be decomposed into bilevel problem as follows:

$$\min_{u,l} \quad l_1 - 2{u_1}^3 + 4l_2$$

s.t.

$$l \in \operatorname{argmin} \left\{ \begin{array}{l} l_1 - 2{u_1}^3 + 4l_2 \\ \text{s.t.} \quad -l_1 - l_2 \leq -5; \quad {u_1}^2 + l_2 \leq 4; \\ \qquad\quad 0 \leq l_1, l_2 \leq 10 \end{array} \right\}$$

$$0 \leq u_1 \leq 10$$

Next, we focus on the approach considered for solving the derived bilevel optimization problem. Solving an optimization problem simply means determining the values of decision variables such that all the constraints remain satisfied and the maximum/minimum objective function value can be attained when the obtained values of decision variables are inserted into constraints and objective function. Since BOBD method classifies the complexity-causing variables into upper level and complexity-soothing variables into lower level, the values of upper level variables should be determined by an evolutionary approach as it can effectively handle the non-regularities introduced by complexity-causing variables and a classical approach should be considered to obtain the values of complexity-soothing lower level variables. Therefore, BOBD method follows hybrid approach to solve the bilevel problem. It uses genetic algorithm to sample the values of upper level variables u and solves the lower level optimization problem using mathematical programming-based method to obtain the values of lower level variables l for any given u. The stepwise procedure of BOBD method is mentioned in Algorithm 1.

Algorithm 1 Bilevel Optimization-Based Decomposition (BOBD)

Input: $F(x)$, $G(x)$- single level optimization problem
Output: x^*- efficient solution of single level optimization problem

Step 1: Classify each variable $x_i \in x$ into upper level (u) or lower level (l) categories (i.e., $x = (u, l)$).

Step 2: Formulate a bilevel problem by decomposing the single level optimization problem with respect to (u, l).

Step 3: Solve the bilevel problem using an evolutionary algorithm at the upper level and a classical algorithm at the lower level (i.e., perform evolutionary sampling of u and use a classical method to solve for l for the given u).

Step 4: Return the best obtained solution ($x^* = (u^*, l^*)$) w.r.t. $F(x)$ and $G(x)$.

To evaluate the performance of the BOBD method, a study by Sinha et al. [240] suggested a test suite of 10 test problems (TP1-TP10). Some of these test problems are from a real-world context. Out of 10 test problems, 8 test problems are scalable in terms of variables and constraints and exhibit various complexities such as nonlinearity, non-convexity, discontinuity, non-differentiability, etc. Authors have shown the effectiveness of BOBD method by solving each test problem using BOBD method, metaheuristic algorithm (GA), and classical methods (interior point and sequential quadratic programming) and later comparing the results from each method in small to large-scale scenarios. The BOBD method outperformed all other methods in small, medium, and large-scale scenarios.

In [240], authors manually classify each variable into upper level (u) or lower level (l) categories (Step 1 in Algorithm 1). In the next study [241], the authors presented an eigen-value and eigen-vector-based variable classification heuristic to automate the variable classification task, which makes BOBD method suitable for large-scale instances by eliminating human intervention. In [242], authors adopted a machine learning approach to automate the variable classification task, wherein a logistic regression model is built to classify the decision variables into

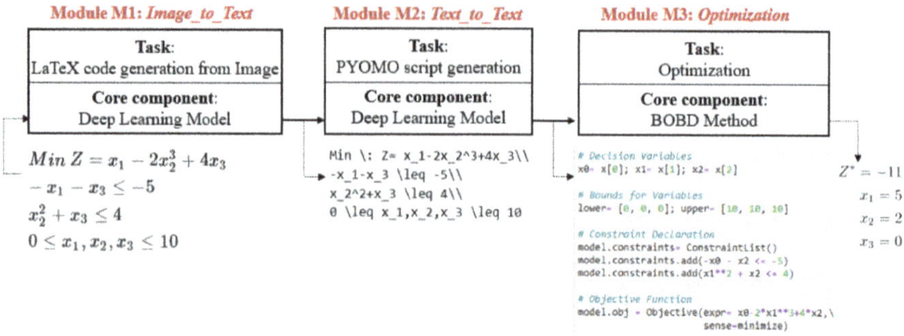

Fig. 3 AutoOpt framework [242] to automate the optimization problem-solving task

upper and lower level categories. In the same study, BOBD method is part of the *AutoOpt* framework designed to automate the optimization problem-solving task, as shown in Fig. 3. *AutoOpt* framework contains three modules in series as follows: M1(*Image_to_Text*)—contains a deep learning model that considers the image of an optimization formulation and generates the corresponding LaTeX code; M2(*Text_to_Text*)—contains a deep learning model that extracts the optimization problem from the LaTeX code and generates a corresponding PYOMO script (programming structure for mathematical modeling language); and M3(*Optimization*)—contains BOBD method to solve the optimization problem from PYOMO script. The procedure followed by *AutoOpt* framework to solve the example discussed in the current section is shown in Fig. 3.

There are several enhancements possible for the BOBD method, for instance, identifying other approaches for variable classification. In BOBD, the complexity-causing variables are classified into upper level and complexity-soothing variables are classified into lower level; hence, lower level optimization problem contains a relatively less complex objective function and constraints. Considering that, several ideas like Ψ-mapping, penalty function, or trust region, can be exploited to reduce the evolutionary sampling at the upper level or the number of calls to the classical algorithm for the lower level.

6.2 *Neural Architecture Search (NAS)*

In machine learning (ML), *architecture parameters* refer to the structural parameters of an algorithm that define overall design and configuration of the algorithm. For example, in an *Artificial Neural Network (ANN)*, the number of layers and the number of neurons per layer constitute key architecture parameters. The process of selecting suitable values for these parameters, known as *architecture engineering*, is carried out before training the model on a given dataset. This step is crucial, as the architecture directly influences the model's learning capacity, generalization ability,

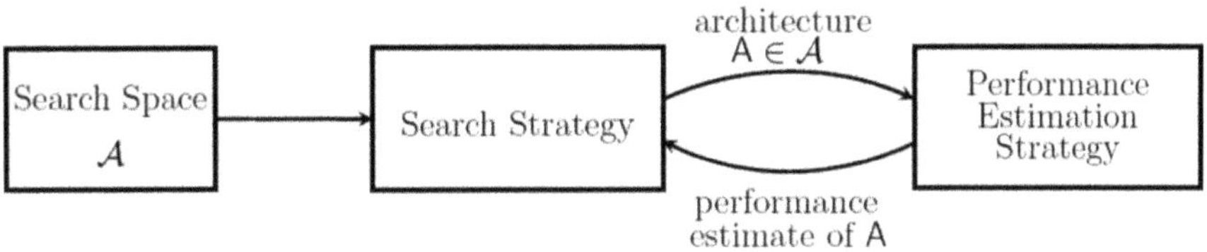

Fig. 4 Outline of NAS: automated process for identifying the optimal architecture from the complex search space

and computational efficiency. Traditionally, architecture engineering was performed manually by human experts, making it both time-consuming and prone to suboptimal choices. Consequently, there has been growing interest in automating this process through *Neural Architecture Search (NAS)*, which systematically explores possible architectures to identify high-performing designs. It is important to note that architecture engineering and NAS are specialized forms of *hyperparameter optimization*, focusing specifically on the structural aspects of machine learning models. NAS methods have already outperformed manually designed architectures on some tasks such as image classification [243, 244], object detection [245], and semantic segmentation [246]. Figure 4 illustrates a general mechanism of NAS [247]. The main aim of NAS is to identify the optimal architecture from a large and complex search space ($\mathcal{A}$) of all possible configurations. Search strategy deals with development of effective technique to explore the search space. Performance estimation strategy dictates how to measure the effectiveness of an architecture ($\mathsf{A} \in \mathcal{A}$) obtained from the search strategy.

In this chapter, we mainly focus on the use of bilevel optimization approach for neural architecture search (NAS). Accordingly, a mathematical formulation representing the implementation of the general mechanism of NAS (Fig. 4) using bilevel optimization approach is provided below:

$$\min_{\mathsf{A}} \quad \mathcal{L}_v(\mathsf{A}, W^*) \tag{42}$$

$$\text{s.t.} \quad W^* \in \operatorname*{argmin}_{W \in \mathcal{W}} \mathcal{L}_t(\mathsf{A}, W) \tag{43}$$

$$\mathsf{A} \in \mathcal{A} \tag{44}$$

In (42–44), the upper level optimization problem aims to minimize the validation loss $\mathcal{L}_v$ w.r.t. the architecture parameters A (upper level variables), and the lower level optimization problem deals with identifying the model parameters W (lower level variables) such that training loss $\mathcal{L}_t$ is minimized for the given A. In this framework, the upper level loop acts as a search strategy whose function is to identify effective architectures by exploring the search space. The lower level problem corresponds to the performance estimation strategy, which involves training a model on a fixed dataset using each architecture proposed by the upper level. There is an assumption that for any architecture A^i passed by the upper level, there exists at least one optimal model parameters $W^*(\mathsf{A}^i)$. This general assumption ensures the well-defined and non-empty feasible region for the upper level problem. Such bilevel approach for NAS enables the joint optimization of architectural choices and model performance. Accordingly, we now discuss several important bilevel approach-based NAS methods, which are broadly classified into two categories: (i) sampling-based NAS methods and (ii) bilevel theory-based methods. The core searching mechanisms of methods from both classes are discussed below.

Sampling-Based NAS Methods: This class of methods uses heuristic or probabilistic sampling strategies to explore the architecture search space. These methods are relatively easy to implement and often serve as a baseline for more sophisticated methods. These methods require a large number of architecture evaluations, which can be computationally expensive. A brief overview of each sampling-based NAS method is provided below.

(1) **Grid Search-Based NAS:** This method follows a deterministic approach that involves trying all possible combinations of architecture parameters and model parameters (i.e., each possible pair of upper and lower level variables (u, l)) defined over a finite and discrete search space.
(2) **Random Search-Based NAS:** This method samples the search space using predefined probability distributions. This stochastic approach often yields better results under a fixed computational budget as compared to the grid search method.
(3) **Bayesian Optimization-Based NAS:** This method uses probabilistic surrogate models to identify the promising regions of the architecture search space. It is more efficient approach compared to grid search and random search approaches. The surrogate models are generally instantiated using Gaussian Processes or Tree-Structured Parzen Estimators (TPE). TPE-based Bayesian optimization method is widely used for NAS. It constructs a probabilistic model that estimates the likelihood of achieving better performance using a given architecture. The acquisition function, part of the TPE-based NAS framework, enables to maintain a balance between exploration and exploitation of search space by strategically selecting the next architecture to evaluate.
(4) **Evolutionary Computation-Based NAS:** Evolutionary computation-based NAS methods use metaheuristic algorithms to effectively explore complex search spaces without relying on gradient information. Though there are methods that integrate evolutionary exploration with hypergradient-based local search [248]. Population-based metaheuristic algorithms, most commonly genetic algorithm or differential evolution, initiate with a population of randomly generated architectures that evolves over iterations as part of the upper level task. At the upper level, crossover and mutation operators are used to update the population or generate new architectures. These architectures are then passed to the lower level, where a model is trained using each architecture and its performance is recorded as the fitness of respective architecture. The latest surveys on the evolutionary computation-based NAS methods are available in [249–251].

(5) **Reinforcement Learning-Based NAS:** This method employs an intelligent agent (typically an RNN controller or Transformer) that learns a policy for sampling efficient architectures through a trial-and-feedback mechanism. In each upper level iteration, the RNN controller, also referred to as RNN-Architecture Sampler, generates a suitable architecture sequentially as per Markov Decision Process (MDP), wherein any architecture A^i is represented as the list of actions $[A_1, A_2, \ldots, A_T]$ (i.e., $A^i = [A_1, A_2, \ldots, A_T]$). The lower level task involves implementing architecture A^i and training the respective model using standard gradient-based method to minimize the training loss. This is a trial part of the trial-and-feedback mechanism. On the feedback side, the performance of trained model is evaluated on the validation set, and the performance result serves as a reward signal for the RNN controller. Later, the reward signal information is used in policy gradient techniques to update the parameters of RNN controller such that the expected reward is maximized. Over the iterations, this procedure gradually refines the sampling distribution and directs the search toward more effective architectures. The recent surveys on the reinforcement learning-based NAS methods (i.e., Automated Reinforcement Learning (AutoRL)) are available in [252, 253].

The studies based on the methods discussed above are provided in Table 2.

Bilevel Theory-Based NAS Methods: This class of methods formulates the search as a nested optimization problem. These methods are mostly based on the classical approach (Sects. 3.1–3.6) and they are built upon formal optimization theories, providing a systematic framework for NAS.

(1) **KKT Reformulation / MPEC-Based NAS**: These methods reformulate the bilevel problem using KKT conditions (16–21) or as a Mathematical Program with Equilibrium Constraints (MPEC) and use appropriate optimization methods to solve the reformulated problem.
(2) **Hypergradient-Based NAS**: These methods compute the hypergradients [278–280], i.e., derivatives of the validation loss w.r.t. architecture parameters, using the reverse mode or implicit differentiation. In some sense, it is similar to the KKT reformulation. DARTS (Differentiable Architecture Search) [281] and its variants fall into this category as it makes the architecture space continuous and differentiable.

Table 2 Search approaches followed in the sampling-based NAS methods

Search approach	Sample studies
Grid search	[254–259]
Random search	[255, 257, 258, 260, 261]
Bayesian optimization-based search	[256, 257, 261–264]
Evolutionary computation-based search	[257, 265–272]
Reinforcement learning-based search	[273–277]

(3) **Penalty-Based NAS**: These methods first reduce the bilevel problem to a single-level formulation and then relax the problem using penalty function or augmented Lagrangian function approach [282, 283]. The optimality of the lower level problem is ensured by adding regularization terms into the upper level objective.
(4) **Surrogate Approximation-Based NAS**: These methods approximate the Ψ-mapping or the φ-mapping using surrogate or meta-models. Approximating the Ψ-mapping emulates the model parameters in response to architecture change [284]. In [282], φ-mapping is approximated using Kriging and then the single-level problem is solved using the penalty-based approach, discussed above.
(5) **Trust Region-Based NAS**: These techniques iteratively solve a series of approximate subproblems within a dynamically updated trust region [285]. These methods ensure stable convergence in scenarios where small changes in hyperparameters lead to significant variations in model performance.

7 Conclusions and Future Research Directions

This chapter presents a comprehensive review of bilevel optimization, covering fundamental principles and solution methods from both classical and evolutionary approaches. The nested structure of bilevel problems, where lower level optimization problem acts as a constraint of upper level optimization problem, often leads to non-convex and disconnected feasible region. These inherent complexities make bilevel optimization a challenging class of optimization problems. The methods from the classical approach are based on the rigorous mathematical optimization theories and they are suitable for bilevel problems following certain regularity conditions (i.e., problems consisting of mathematically well-behaving objective functions and constraints). The core mechanisms of such classical methods, including single-level reduction methods, descent methods, penalty function methods, and trust region methods, are discussed. The evolutionary class of methods employs the nature-inspired or intelligence-based strategies to effectively explore the complex feasible region characterized by nonlinearity, non-convexity, discreteness, discontinuity, etc. The use of evolutionary method along with classical method, a hybrid approach, is also discussed in the context of bilevel problem solving. Later, the mixed-integer and multiobjective scenarios, which need further attention in bilevel optimization field, are also addressed. Two recent applications of bilevel approach, i.e., bilevel optimization-based decomposition and neural architecture search for machine learning algorithms, are discussed in detail. These new applications have the potential to impact the areas of optimization and machine learning in a significant manner. Interestingly, the first application supports the automation of optimization problem solving, and the second application supports the automation of neural architecture design.

Although significant progress has been observed in recent years, bilevel optimization continues to be in a developmental stage, offering substantial scope for both computational and theoretical contributions. Apart from that, discrete and multi-objective bilevel optimization topics have received limited attention in the bilevel research community. From a computational resource perspective, there is potential for using distributed computing platforms to effectively handle large-scale bilevel problems. Bilevel optimization remains an active research area, and the development of increasingly efficient algorithms is driving a shift toward more application-oriented studies. Simultaneously, emerging practical challenges are giving rise to newer types of practical bilevel problems. As a result, a broader spectrum of real-world applications is expected to surface in the near future.

References

1. Sinha, A., Malo, P., Deb, K.: A review on bilevel optimization: from classical to evolutionary approaches and applications. IEEE Trans. Evol. Comput. **22**(2), 276–295 (2017)
2. Dempe, S.: Foundations of Bilevel Programming. Kluwer Academic Publishers, Secaucus, NJ, USA (2002)
3. Ryu, J.-H., Dua, V., Pistikopoulos, E.N.: A bilevel programming framework for enterprise-wide process networks under uncertainty. Comput. Chem. Eng. **28**(6–7), 1121–1129 (2004)
4. Sinha, A., Malo, P., Deb, K., Korhonen, P., Wallenius, J.: Solving bilevel multicriterion optimization problems with lower level decision uncertainty. IEEE Trans. Evol. Comput. **20**(2), 199–217 (2015)
5. Burtscheidt, J., Claus, M.: Bilevel linear optimization under uncertainty. In: Bilevel Optimization: Advances and next Challenges, pp. 485–511 (2020)
6. Beck, Y., Ljubić, I., Schmidt, M.: A survey on bilevel optimization under uncertainty. Eur. J. Oper. Res. **311**(2), 401–426 (2023)
7. Zhichao, L., Deb, K., Sinha, A.: Finding reliable solutions in bilevel optimization problems under uncertainties. In: Proceedings of the Genetic and Evolutionary Computation Conference **2016**, 941–948 (2016)
8. Zhichao, L., Deb, K., Sinha, A.: Uncertainty handling in bilevel optimization for robust and reliable solutions. Internat. J. Uncertain. Fuzziness Knowl.-Based Syst. **26**(Suppl. 2), 1–24 (2018)
9. Bhatt, S., Sinha, A., Jayaswal, S.: Protection of capacitated hubs under demand uncertainty: a robust optimization approach. Working paper, Indian Institute of Management Ahmedabad (2026)
10. Lozano, L., Borrero, J.S.: A bilevel optimization approach for a class of combinatorial problems with disruptions and probing. INFORMS J, Comput (2024)
11. Bostian, M., Sinha, A., Gerald, W., Bradley, B.: Incorporating data envelopment analysis solution methods into bilevel multi-objective optimization. In: 2015 IEEE Congress on Evolutionary Computation (CEC-2015). IEEE Press (2015)
12. Whittaker, G., Färe, R., Grosskopf, S., Barnhart, B., Bostian, M., Mueller-Warrant, G., Griffith, S.: Spatial targeting of agri-environmental policy using bilevel evolutionary optimization. Omega (2016)
13. Barnhart, B., Lu, Z., Bostian, M., Sinha, A., Deb, K., Kurkalova, L., Jha, M., Whittaker, G.: Handling practicalities in agricultural policy optimization for water quality improvements. In: Proceedings of the Genetic and Evolutionary Computation Conference, pp. 1065–1072 (2017)

14. Brown, G., Matthew, C., Douglas, D., Jeffrey, K., Wood, K.: A Two-sided optimization for theater ballistic missile defense. Oper. Res. **53**(5), 745–763 (2005)
15. Fischetti, M., Ljubić, I., Monaci, M., Sinnl, M.: Interdiction games and monotonicity, with application to knapsack problems. INFORMS J. Comput. **31**(2), 390–410 (2019)
16. Brown, G.G., Carlyle, W.M., Harney, R.C., Skroch, E.M., Wood, K.: Interdicting a nuclear-weapons project. Oper. Res. **57**(4), 866–877 (2009)
17. Bucarey, V., Casorrán, C., Labbé, M., Ordoñez, F., Figueroa, O.: Coordinating resources in stackelberg security games. Eur. J. Oper. Res. **291**(3), 846–861 (2021)
18. Casorrán, C., Fortz, B., Labbé, M., Ordóñez, F.: A study of general and security Stackelberg game formulations. Eur. J. Oper. Res. **278**(3), 855–868 (2019)
19. Ordónez, F., Tambe, M., Jara, J.F., Jain, M., Kiekintveld, C., Tsai, J.: Deployed security games for patrol planning. In: Handbook of Operations Research for Homeland Security, pp. 45–72. Springer (2012)
20. Shan, X., Zhuang, J.: Hybrid defensive resource allocations in the face of partially strategic attackers in a sequential defender-attacker game. Eur. J. Oper. Res. **228**(1), 262–272 (2013)
21. Baggio, A., Carvalho, M., Lodi, A., Tramontani, A.: Multilevel approaches for the critical node problem. Oper. Res. **69**(2), 486–508 (2021)
22. Wein, L.M.: Homeland security: from mathematical models to policy implementation: the 2008 philip McCord morse lecture. Oper. Res. **57**(4), 801–811 (2009)
23. An, B., Ordóñez, F., Tambe, M., Shieh, E., Yang, R., Baldwin, C., DiRenzo, J., III., Moretti, K., Maule, B., Meyer, G.: A deployed quantal response-based patrol planning system for the us coast guard. Interfaces **43**(5), 400–420 (2013)
24. Furini, F., Ljubić, I., Martin, S., Segundo, P.S.: The maximum clique interdiction problem. Eur. J. Operat. Res. **277**(1), 112–127 (2019)
25. Bennett, K.P., Kunapuli, G., Hu, J., Pang, J.-S.: Bilevel optimization and machine learning. In: Computational Intelligence: Research Frontiers, pp. 25–47. Springer (2008)
26. Sinha, A., Khandait, T., Mohanty, R.: A gradient-based bilevel optimization approach for tuning hyperparameters in machine learning. ArXiv preprint arXiv:2007.11022 (2020)
27. Sinha, A., Malo, P., Peng, X., Deb, K.: A bilevel optimization approach to automated parameter tuning. In: Proceedings of the 16th Annual Genetic and Evolutionary Computation Conference (GECCO 2014), pp. 847–854. ACM Press, New York (2014)
28. Liu, B., Ye, M., Wright, S., Stone, P., Liu, Q.: Bome! bilevel optimization made easy: a simple first-order approach. Adv. Neural. Inf. Process. Syst. **35**, 17248–17262 (2022)
29. Petrulionytė, I., Mairal, J., Arbel, M.: Functional bilevel optimization for machine learning. Adv. Neural. Inf. Process. Syst. **37**, 14016–14065 (2024)
30. Hansen, P., Jaumard, B., Savard, G.: New branch-and-bound rules for linear bilevel programming. SIAM J. Sci. Stat. Comput. **13**(5), 1194–1217 (1992)
31. Vicente, L., Savard, G., Joaquim, J.: Descent approaches for quadratic bilevel programming. J. Optim. Theory Appl. **81**(2), 379–399 (1994)
32. Deng, X.: Complexity issues in bilevel linear programming. In: Multilevel Optimization: Algorithms and Applications, pp. 149–164. Springer (1998)
33. Deb, K., Sinha, A.: An efficient and accurate solution methodology for bilevel multi-objective programming problems using a hybrid evolutionary-local-search algorithm. Evol. Comput. J. **18**(3), 403–449 (2010)
34. Sinha, A., Malo, P., Deb, K.: Efficient evolutionary algorithm for single-objective bilevel optimization. ArXiv preprint arXiv:1303.3901, 2013
35. Angelo, J.S., Eduardo, K., Barbosa Helio J.C.: Differential evolution for bilevel programming. In: Proceedings of the 2013 Congress on Evolutionary Computation (CEC-2013). IEEE Press (2013)
36. Sinha, A., Malo, P., Deb, K.: Evolutionary algorithm for bilevel optimization using approximations of the lower level optimal solution mapping. Eur. J. Oper. Res. **257**(2), 395–411 (2017)
37. Ankur Sinha, L., Zhichao, K.D., Malo, P.: Bilevel optimization based on iterative approximation of multiple mappings. J. Heuristics **26**(2), 151–185 (2020)

38. Dempe, S., Joydeep, D., Mordukhovich, B.S.: New necessary optimality conditions in optimistic bilevel programming. Optimization **56**(5–6), 577–604 (2007)
39. Harker, P.T., Pang, J.-S.: Existence of optimal solutions to mathematical programs with equilibrium constraints. Oper. Res. Lett. **7**, 61–64 (1988)
40. Lignola, M.B., Morgan, J.: Topological existence and stability for Stackelberg problems. J. Optim. Theory Appl. **84**, 145–169 (1995)
41. Lignola, M.B., Morgan, J.: Existence of solutions to bilevel variational problems in banach spaces. In: Equilibrium Problems: Nonsmooth Optimization and Variational Inequality Models, pp. 161–174. Springer (2001)
42. Outrata, J.V.: Necessary optimality conditions for Stackelberg problems. J. Optim. Theory Appl. **76**, 305–320 (1993)
43. Dempe, S., Franke, S.: Solution of bilevel optimization problems using the kkt approach. Optimization **68**(8), 1471–1489 (2019)
44. Dempe, S., Mordukhovich, B.S., Zemkoho, A.B.: Two-level value function approach to non-smooth optimistic and pessimistic bilevel programs. Optimization **68**(2–3), 433–455 (2019)
45. Antoniou, M., Sinha, A., Papa, G.: δ-perturbation of bilevel optimization problems: an error bound analysis. Oper. Res. Perspect. **13**, 100315 (2024)
46. Dempe, S., Mordukhovich, B.S., Zemkoho, A.B.: Necessary optimality conditions in pessimistic bilevel programming. Optimization **63**(4), 505–533 (2014)
47. Loridan, P., Morgan, J.: Weak via strong Stackelberg problem: new results. J. Global Optim. **8**, 263–287 (1996)
48. Lucchetti, R., Mignanego, F., Pieri, G.: Existence theorem of equilibrium points in Stackelberg games with constraints. Optimization **18**, 857–866 (1987)
49. Wiesemann, W., Tsoukalas, A., Kleniati, P.-M., Rustem, B.: Pessimistic bilevel optimization. SIAM J. Optim. **23**(1), 353–380 (2013)
50. Antoniou, M., Sinha, A., Papa, G.: A study on optimistic and pessimistic pareto-fronts in multi-objective bilevel optimization via δ-perturbation. In: International Conference on Evolutionary Multi-Criterion Optimization, pp. 103–117. Springer (2025)
51. Liu, J., Fan, Y., Chen, Z., Zheng, Y.: Methods for pessimistic bilevel optimization. In: Bilevel Optimization: Advances and Next Challenges, pp. 403–420 (2020)
52. Kleinert, T., Labbé, M., Ljubić, I., Schmidt, M.: A survey on mixed-integer programming techniques in bilevel optimization. EURO J. Comput. Optim. **9**, 100007 (2021)
53. Calvete, H.I., Galé, C.: Algorithms for linear bilevel optimization. In: Bilevel Optimization: Advances and Next Challenges, pp. 293–312. Springer (2020)
54. Smith, J.C., Song, Y.: A survey of network interdiction models and algorithms. Eur. J. Oper. Res. **283**(3), 797–811 (2020)
55. Sinha, A., Malo, P., Deb, K.: Evolutionary bilevel optimization: an introduction and recent advances. In: Recent Advances in Evolutionary Multi-objective Optimization, pp. 71–103 (2016)
56. Dempe, S.: Bilevel optimization: theory, algorithms, applications and a bibliography. In: Bilevel Optimization: Advances and Next Challenges, pp. 581–672. Springer (2020)
57. Sinha, A., Deb, K.: Bilevel multi-objective optimization and decision making. In: Metaheuristics for Bi-level Optimization, pp. 247–284. Springer (2013)
58. Colson, B., Patrice, M., Savard, G.: An overview of bilevel optimization. Ann. Oper. Res. **153**, 235–256 (2007)
59. Dempe, S.: Annotated bibliography on bilevel programming and mathematical programs with equilibrium constraints. Optimization **52**(3), 339–359 (2003)
60. Kalashnikov, V.V., Dempe, S., Pérez-Valdés, G.A., Kalashnykova, N.I., Camacho-Vallejo, J.-F.: Bilevel programming and applications. Math. Probl, Eng (2015)
61. Vicente, L.N., Bilevel, C.P.H.: multilevel programming: a bibliography review. J. Glob. Optim. **5**, 291–306 (1994)
62. Bard, J.F.: Practical Bilevel Optimization: Algorithms and Applications. Kluwer, The Netherlands (1998)

63. Colson, B.: Mathematical programs with equilibrium constraints and nonlinear bilevel programming problems. Technical report, Master's thesis, Department of Mathematics, FUNDP, Namur, Belgium (1999)
64. Shimizu, K., Yo, I., Bard, J.F.: Nondifferentiable and Two-Level Mathematical Programming. Kluwer Academic, Dordrecht (1997)
65. Aksen, D., Aras, N.: A matheuristic for leader-follower games involving facility location-protection-interdiction decisions. In: Metaheuristics for Bi-Level Optimization, PP. 115–151. Springer (2013)
66. Talbi, E.-G.: *Metaheuristics for Bi-Level Optimization*, VOL. 482. Springer (2013)
67. Dempe, S., Kalashnikov, V., Pérez-Valdés, G.A., Kalashnykova, N.: Theory, algorithms and applications to energy networks. Energy Systems. Springer, Berlin, Bilevel Programming Problems (2015)
68. Zemkoho, A., Dempe, S.: Bilevel optimization advances and next challenges (2020)
69. Dempe, S., Zemkoho, A.: Bilevel optimization. In *Springer Optimization and Its Applications*, volume 161. Springer (2020)
70. Dempe, S., Dutta, J.: Is bilevel programming a special case of a mathematical program with complementarity constraints? Math. Program. **131**(1), 37–48 (2012)
71. Bialas, W.F., Karwan, M.H.: Two-level linear programming. Manage. Sci. **30**(8), 1004–1020 (1984)
72. Chen, Y., Florian, M.: On the geometry structure of linear bilevel programs: a dual approach. Technical Report CRT-867, Centre de Recherche sur les Transports (1992)
73. Tuy, H., Athanasios, M., Peter, V.: A global optimization approach for the linear two-level program. J. Global Optim. **3**, 1–23 (1993)
74. Fortuny-Amat, J., Bruce, M.C.: A representation and economic interpretation of a two-level programming problem. J. Operat. Res. Soc. **32**, 783–792 (1981)
75. Bard, J.F., Falk, J.E.: An explicit solution to the multi-level programming problem. Comput. Oper. Res. **9**, 77–100 (1982)
76. Bard, J.F., Moore, J.T.: A branch and bound algorithm for the bilevel programming problem. SIAM J. Sci. Stat. Comput. **11**, 281–292 (1990)
77. Al-Khayyal, F.A., Reiner, H., Pardalos, P.M.: Global optimization of concave functions subject to quadratic constraints: an application in nonlinear bilevel programming. Ann. Oper. Res. **34**, 125–147 (1992)
78. Edmunds, T.A., Bard, J.F.: Algorithms for nonlinear bilevel mathematical programming. IEEE Trans. Syst. Man Cybernet. **21**, 83–89 (1991)
79. Shi, C., Jie, L., Zhang, G.: An extended kuhn-tucker approach for linear bilevel programming. Appl. Math. Comput. **162**(1), 51–63 (2005)
80. Sinha, A., Soun, T., Deb, K.: Using karush-kuhn-tucker proximity measure for solving bilevel optimization problems. Swarm Evol. Comput. **44**, 496–510 (2019)
81. Jayaswal, S., Sinha, A.: Bilevel optimization: applications, models and solution approaches. In: Optimization Essentials: Theory, Tools, and Applications, pp. 469–499. Springer (2024)
82. Dempe, S., Franke, S.: On the solution of convex bilevel optimization problems. Comput. Optim. Appl. **63**(3), 685–703 (2016)
83. Lozano, L., Smith, J.C.: A value-function-based exact approach for the bilevel mixed-integer programming problem. Operat. Res. **65**(3), 768–786 (2017)
84. Sinha, A., Bedi, S., Deb, K.: Bilevel optimization based on kriging approximations of lower level optimal value function. In: 2018 IEEE Congress on Evolutionary Computation (CEC), pp. 1–8. IEEE (2018)
85. Liu, R., Liu, X., Yuan, X., Zeng, S., Zhang, J.: A value-function-based interior-point method for non-convex bi-level optimization. In: International Conference on Machine Learning, pp. 6882–6892. PMLR (2021)
86. Kolstad, C.D., Lasdon, L.S.: Derivative evaluation and computational experience with large bilevel mathematical programs. J. Optim. Theory Appl. **65**, 485–499 (1990)
87. Savard, G., Gauvin, J.: The steepest descent direction for the nonlinear bilevel programming problem. Oper. Res. Lett. **15**, 275–282 (1994)

88. Vicente, L., Savard, G., Joaquim, J.: Descent approaches for quadratic bilevel programming. J. Optim. Theory Appl. **81**, 379–399 (1994)
89. Aiyoshi, E., Kiyotaka, S.: Hierarchical decentralized systems and its new solution by a barrier method. IEEE Trans. Syst. Man Cybern. **6**, 444–449 (1981)
90. Aiyoshi, E., Kiyotaka, S.: A solution method for the static constrained Stackelberg problem via penalty method. IEEE Trans. Autom. Control **29**, 1111–1114 (1984)
91. Yo, I., Aiyoshi, E.: Double penalty method for bilevel optimization problems. Ann. Oper. Res. **34**, 73–88 (1992)
92. White, D.J., Anandalingam, G.: A penalty function approach for solving bi-level linear programs. J. Global Optim. **3**, 397–419 (1993)
93. Lv, Y., Tiesong, H., Wang, G., Wan, Z.: A penalty function method based on kuhn-tucker condition for solving linear bilevel programming. Appl. Math. Comput. **188**(1), 808–813 (2007)
94. Liu, G., Jiye, H., Shouyang, W.: A trust region algorithm for bilevel programing problems. Chin. Sci. Bull. **43**(10), 820–824 (1998)
95. Marcotte, P., Savard, G., Zhu, D.L.: A trust region algorithm for nonlinear bilevel programming. Oper. Res. Lett. **29**(4), 171–179 (2001)
96. Colson, B., Patrice, M., Savard, G.: A trust-region method for nonlinear bilevel programming: algorithm and computational experience. Comput. Optim. Appl. **30**(3), 211–227 (2005)
97. Li, H., Yuping, W.: A hybrid genetic algorithm for solving nonlinear bilevel programming problems based on the simplex method. Int. Conf. Nat. Comput. **4**, 91–95 (2007)
98. Mathieu, R., Pittard, L., Anandalingam, G.: Genetic algorithm based approach to bi-level linear programming. Oper. Res. **28**(1), 1–21 (1994)
99. Yafeng, Y.: Genetic algorithm based approach for bilevel programming models. J. Transp. Eng. **126**(2), 115–120 (2000)
100. Zhu, X. Qian, Y., Xianjia, W.: A hybrid differential evolution algorithm for solving nonlinear bilevel programming with linear constraints. In: 5th IEEE International Conference on Cognitive Informatics, 2006. ICCI 2006. , volume 1, pp. 126–131. IEEE (2006)
101. Reza, H.S., Azizollah, M., Jahanshahloo, G., Mohammad Mehdi, S.: Linear bilevel programming solution by genetic algorithm. Comput. Operat. Res. **29**(13), 1913–1925 (2002)
102. Calvete, H.I., Galé, C., Oliveros, M.-J.: Bilevel model for production-distribution planning solved by using ant colony optimization. Comput. Operat. Res. **38**(1), 320–327 (2011)
103. Li, H.: A genetic algorithm using a finite search space for solving nonlinear/linear fractional bilevel programming problems. Ann. Operat. Res. 1–16 (2015)
104. Wang, Y., Yong-Chang, J., Hong, L.: An evolutionary algorithm for solving nonlinear bilevel programming based on a new constraint-handling scheme. IEEE Trans. Syst. Man, Cybern., Part C (Appl. Rev.) **35**(2), 221–232 (2005)
105. Wang, Y., Hong, L., Chuangyin, D.: A new evolutionary algorithm for a class of nonlinear bilevel programming problems and its global convergence. INFORMS J. Comput. **23**(4), 618–629 (2011)
106. Li, X., Peng, T., Xiaoping, M.: A hierarchical particle swarm optimization for solving bilevel programming problems. In: Artificial Intelligence and Soft Computing—ICAISC 2006, vol. 4029, pp. 1169–1178. Springer, Berlin Heidelberg (2006)
107. Jiang, Y., Li, X., Huang, C., Xianing, W.: Application of particle swarm optimization based on chks smoothing function for solving nonlinear bilevel programming problem. Appl. Math. Comput. **219**(9), 4332–4339 (2013)
108. Wan, Z., Wang, G., Sun, B.: A hybrid intelligent algorithm by combining particle swarm optimization with chaos searching technique for solving nonlinear bilevel programming problems. Swarm Evol. Comput. **8**, 26–32 (2013)
109. Angelo, J.S., Eduardo, K., Helio, B.J.C.: Differential evolution assisted by a surrogate model for bilevel programming problems. In: 2014 IEEE Congress on Evolutionary Computation (CEC), pp. 1784–1791. IEEE (2014)
110. Angelo, J.S., Eduardo, K., Barbosa Helio J.C.: Differential evolution for bilevel programming. In: 2013 IEEE Congress on Evolutionary Computation, pp. 470–477. IEEE (2013)

111. Camacho-Vallejo, J.-F., Muñoz-Sánchez, R., González-Velarde, J.L.: A heuristic algorithm for a supply chain's production-distribution planning. Comput. Operat. Res. **61**, 110–121 (2015)
112. Camacho-Vallejo, J.-F., Corpus, C., Villegas, J.G.: Metaheuristics for bilevel optimization: a comprehensive review. Comput. Operat. Res. **161**, 106410 (2024)
113. Deb, K., Sinha, A., Malo, P., Lu, Z.: Approximate bilevel optimization with population-based evolutionary algorithms. In: Bilevel Optimization: Advances and Next Challenges, pp. 361–402 (2020)
114. Sinha, A., Malo, P., Frantsev, A., Deb, K.: Finding optimal strategies in a multi-period multi-leader-follower stackelberg game using an evolutionary algorithm. Comput. Operat. Res. **41**, 374–385 (2014)
115. Angelo, J.S., Helio B.J.C.: A study on the use of heuristics to solve a bilevel programming problem. Int. Trans. Operat. Res. (2015)
116. Islam, M.M., Kumar Singh, H., Ray, T., Sinha, A.: An enhanced memetic algorithm for single-objective bilevel optimization problems. Evolutionary Comput. **25**(4), 607–642 (2017)
117. Wang, G., Wan, Z., Wang, X., Lv, Y.: Genetic algorithm based on simplex method for solving linear-quadratic bilevel programming problem. Comput. Math. Appl. **56**(10), 2550–2555 (2008)
118. Sinha, A., Malo, P., Deb, K.: An improved bilevel evolutionary algorithm based on quadratic approximations. In: 2014 IEEE Congress on Evolutionary Computation (CEC), pp. 1870–1877. IEEE (2014)
119. Sinha, A., Malo, P., Deb, K.: Solving optimistic bilevel programs by iteratively approximating lower level optimal value function. In: 2016 IEEE Congress on Evolutionary Computation (CEC), pp. 1877–1884. IEEE (2016)
120. Islam, M.M., Singh, H.K., Ray, T.: A surrogate assisted approach for single-objective bilevel optimization. IEEE Trans. Evolut. Comput. **21**(5), 681–696 (2017)
121. Vicente, L., Savard, G., Judice, J.: Discrete linear bilevel programming problem. J. Optim. Theory Appl. **89**(3), 597–614 (1996)
122. Moore, J.T., Bard, J.F.: The mixed integer linear bilevel programming problem. Oper. Res. **38**(5), 911–921 (1990)
123. Bard, J.F., Moore, J.T.: An algorithm for the discrete bilevel programming problem. Nav. Res. Logist. (NRL) **39**(3), 419–435 (1992)
124. Dempe, S.: Discrete Bilevel Optimization Problems. Citeseer (1996)
125. Caramia, M., Mari, R.: A decomposition approach to solve a bilevel capacitated facility location problem with equity constraints. Optim. Lett., 1–23 (2015)
126. Fontaine, P., Minner, S.: Benders decomposition for discrete-continuous linear bilevel problems with application to traffic network design. Transp. Res. Part B: Methodol. **70**, 163–172 (2014)
127. Saharidis, G.K., Ierapetritou, M.G.: Resolution method for mixed integer bi-level linear problems based on decomposition technique. J. Global Optim. **44**(1), 29–51 (2009)
128. Fischetti, M., Ljubić, I., Monaci, M., Sinnl, M.: A new general-purpose algorithm for mixed-integer bilevel linear programs. Oper. Res. **65**(6), 1615–1637 (2017)
129. Fischetti, M., Ljubić, I., Monaci, M., Sinnl, M.: On the use of intersection cuts for bilevel optimization. Math. Program. **172**(1), 77–103 (2018)
130. Yue, D., Gao, J., Zeng, B., You, F.: A projection-based reformulation and decomposition algorithm for global optimization of a class of mixed integer bilevel linear programs. J. Global Optim. **73**(1), 27–57 (2019)
131. Tahernejad, S., Ralphs, T.K., DeNegre, S.T.: A branch-and-cut algorithm for mixed integer bilevel linear optimization problems and its implementation. Math. Program. Comput. **12**(4), 529–568 (2020)
132. DeNegre, S.T., Ralphs, T.K.: A branch-and-cut algorithm for integer bilevel linear programs. In: Operations Research and Cyber-Infrastructure, pp. 65–78. Springer (2009)
133. Grimm, V., Kleinert, T., Liers, F., Schmidt, M., Zöttl, G.: Optimal price zones of electricity markets: a mixed-integer multilevel model and global solution approaches. Optimization methods and software **34**(2), 406–436 (2019)

134. Arroyo, J.M., Federico, F.J.: A genetic algorithm approach for the analysis of electric grid interdiction with line switching. In: 15th International Conference on Intelligent System Applications to Power Systems. ISAP'09, pp. 1–6. IEEE (2009)
135. J.-F. Camacho-Vallejo, Cordero-Franco, Á.E., González-Ramírez, R.G.: Solving the bilevel facility location problem under preferences by a Stackelberg-evolutionary algorithm. Math. Probl. Eng. (2014)
136. Chaabani, A., Bechikh, S., Said, L.B.: A co-evolutionary decomposition-based algorithm for bi-level combinatorial optimization. In: 2015 IEEE Congress on Evolutionary Computation (CEC), pp. 1659–1666. IEEE (2015)
137. Daniel Handoko, S., Chuin, L.H., Gupta, A., Soon, O.Y., Kim, H.C., Siew, T.P.: Solving multi-vehicle profitable tour problem via knowledge adoption in evolutionary bi-level programming. In: 2015 IEEE Congress on Evolutionary Computation (CEC), pp. 2713–2720. IEEE (2015)
138. Hecheng, L., Yuping, W.: Exponential distribution-based genetic algorithm for solving mixed-integer bilevel programming problems. J. Syst. Eng. Electron. **19**(6), 1157–1164 (2008)
139. Legillon, F., Liefooghe, A., Talbi, E.-G.: Cobra: a cooperative coevolutionary algorithm for bi-level optimization. In: 2012 IEEE Congress on Evolutionary Computation, pp. 1–8. IEEE (2012)
140. Miandoabchi, E., Farahani, R.Z.: Optimizing reserve capacity of urban road networks in a discrete network design problem. Adv. Eng. Softw. **42**(12), 1041–1050 (2011)
141. Kleinert, T., Schmidt, M.: Computing feasible points of bilevel problems with a penalty alternating direction method. INFORMS J. Comput. **33**(1), 198–215 (2021)
142. Kleinert, T., Schmidt, M.: Global optimization of multilevel electricity market models including network design and graph partitioning. Discret. Optim. **33**, 43–69 (2019)
143. Avraamidou, S., Pistikopoulos, E.N.: A multi-parametric optimization approach for bilevel mixed-integer linear and quadratic programming problems. Comput. Chem. Eng. **125**, 98–113 (2019)
144. Kleinert, T., Labbé, M., Plein, F., Schmidt, M.: Closing the gap in linear bilevel optimization: a new valid primal-dual inequality. Optim. Lett. **15**(4), 1027–1040 (2021)
145. Liu, S., Wang, M., Kong, N., Xiangpei, H.: An enhanced branch-and-bound algorithm for bilevel integer linear programming. Eur. J. Oper. Res. **291**(2), 661–679 (2021)
146. Gadhi, N., Dempe, S.: Necessary optimality conditions and a new approach to multiobjective bilevel optimization problems. J. Optim. Theory Appl. **155**(1), 100–114 (2012)
147. Jane, Y.J.: Necessary optimality conditions for multiobjective bilevel programs. Math. Oper. Res. **36**(1), 165–184 (2011)
148. Bank, B., Guddat, J., Kummer, B., Tammer, K.: Non-linear Parametric Optimization. Birkhäuser Basel, Diethard Klatte (1983)
149. Shi, X., Sheng, X.H.: Model and interactive algorithm of bi-level multi-objective decision-making with multiple interconnected decision makers. J. Multi-Criteria Decis. Anal. **10**(1), 27–34 (2001)
150. Eichfelder, G.: Solving nonlinear multiobjective bilevel optimization problems with coupled upper level constraints. Technical Report Preprint No. 320, Preprint-Series of the Institute of Applied Mathematics, Univ. Erlangen-Nornberg, Germany (2007)
151. Eichfelder, G.: Multiobjective bilevel optimization. Math. Program. **123**(2), 419–449 (2010)
152. Lv, Y., Wan, Z.: A solution method for the optimistic linear semivectorial bilevel optimization problem. J. Inequalities Appl. **2014**(1), 164 (2014)
153. Gupta, A., Ong, Y.-S.: An evolutionary algorithm with adaptive scalarization for multiobjective bilevel programs. In: 2015 IEEE Congress on Evolutionary Computation (CEC), pp. 1636–1642. IEEE (2015)
154. Bonnel, H.: Optimality conditions for the semivectorial bilevel optimization problem. Pacific J. Opt. **2** (2006)
155. Ankhili, Z., Mansouri, A.: An exact penalty on bilevel programs with linear vector optimization lower level. Eur. J. Oper. Res. **197**(1), 36–41 (2009)
156. Zheng, Y., Wan, Z.: A solution method for semivectorial bilevel programming problem via penalty method. J. Appl. Math. Comput. **37**(1), 207–219 (2011)

157. Calvete, H.I., Galé, C.: On linear bilevel problems with multiple objectives at the lower level. Omega **39**(1), 33–40 (2011)
158. Ren, A., Wang, Y.: A novel penalty function method for semivectorial bilevel programming problem. Appl. Math. Model. **40**(1), 135–149 (2016)
159. Halter, W., Sanaz, M.: Bilevel optimization of multi-component chemical systems using particle swarm optimization. In: Proceedings of World Congress on Computational Intelligence (WCCI-2006), pp. 1240–1247 (2006)
160. Sinha, A., Deb, K.: Towards understanding evolutionary bilevel multi-objective optimization algorithm. IFAC Proc. Volumes **42**(2), 338–343 (2009)
161. Sinha, A.: Bilevel multi-objective optimization problem solving using progressively interactive emo. In: International Conference on Evolutionary Multi-Criterion Optimization, pp. 269–284. Springer (2011)
162. Sinha, A., Malo, P., Deb, K.: Towards understanding bilevel multi-objective optimization with deterministic lower level decisions. In: International Conference on Evolutionary Multi-criterion Optimization, pp. 426–443. Springer (2015)
163. Deb, K., Sinha, A.: Constructing test problems for bilevel evolutionary multi-objective optimization. In: 2009 IEEE Congress on Evolutionary Computation (CEC-2009), pp. 1153–1160. IEEE Press (2009)
164. Deb, K., Sinha, A.: An evolutionary approach for bilevel multi-objective problems. In: Cutting-Edge Research Topics on Multiple Criteria Decision Making. Communications in Computer and Information Science, vol. 35, pp. 17–24. Springer, Berlin, Germany (2009)
165. Sinha, A., Malo, P., Deb, K.: Approximated set-valued mapping approach for handling multiobjective bilevel problems. Comput. Operat. Res. **77**, 194–209 (2017)
166. Deb, K., Lu, Z., Kropp, I., Hernandez-Suarez, J.S., Hussein, R., Miller, S., Nejadhashemi, A.P.: Minimizing expected deviation in upper level outcomes due to lower level decision making in hierarchical multiobjective problems. IEEE Trans. Evolut. Comput. **27**(3), 505–519 (2022)
167. Alves, M.J., Antunes, C.H., Costa, J.P.: New concepts and an algorithm for multiobjective bilevel programming: optimistic, pessimistic and moderate solutions. Oper. Res. **21**(4), 2593–2626 (2021)
168. Mejía-De-Dios, J.-A., Rodríguez-Molina, A., Mezura-Montes, E.: Multiobjective bilevel optimization: a survey of the state-of-the-art. IEEE Trans. Syst., Man, Cybern.: Syst. **53**(9), 5478–5490 (2023)
169. Linnala, M., Madetoja, E., Ruotsalainen, H., Hämäläinen, J.: Bi-level optimization for a dynamic multiobjective problem. Eng. Optim. **44**(2), 195–207 (2012)
170. Pieume, C.O., Fotso, L.P., Siarry, P.: Solving bilevel programming problems with multicriteria optimization techniques. Opsearch **46**(2), 169–183 (2009)
171. Pramanik, S., Dey, P.P.: Bi-level multi-objective programming problem with fuzzy parameters. Int. J. Comput. Appl. **30**(10), 13–20, September 2011. Published by Foundation of Computer Science, New York, USA
172. Ruuska, S., Kaisa, M.: Constructing evolutionary algorithms for bilevel multiobjective optimization. In: 2012 IEEE Congress on Evolutionary Computation (CEC), pp. 1–7 (2012)
173. Zhang, T., Tiesong, H., Yue, Z., Xuning, G.: An improved particle swarm optimization for solving bilevel multiobjective programming problem. J. Appl, Math (2012)
174. Juan Zou, X., Yang, Z.L., Liu, J., Zhang, L., Zheng, J.: Multiobjective bilevel optimization algorithm based on preference selection to solve energy hub system planning problems. Energy **232**, 120995 (2021)
175. Chen, J., Ding, J., Li, K., Tan, K.C., Chai, T.: A knee point driven evolutionary algorithm for multiobjective bilevel optimization. IEEE Trans. Cybern. **54**(7), 4177–4189 (2024)
176. Wang, J.Y.T., Matthias, E., Dirks Kim N., Gupta, A.: A bilevel multi-objective road pricing model for economic, environmental and health sustainability. Transp. Res. Proced. **3**, 393–402 (2014)
177. Sinha, A., Malo, P., Deb, K.: Transportation policy formulation as a multi-objective bilevel optimization problem. In: 2015 IEEE Congress on Evolutionary Computation (CEC-2015). IEEE Press (2015)

178. Migdalas, A.: Bilevel programming in traffic planning: models, methods and challenge. J. Global Optim. **7**(4), 381–405 (1995)
179. Marcotte, P., Savard, G., Semet, F.: A bilevel programming approach to the travelling salesman problem. Oper. Res. Lett. **32**(3), 240–248 (2004)
180. Labbé, M., Patrice, M., Savard, G.: A Bilevel Model of Taxation and Its Application to Optimal Highway Pricing. Manage. Sci. **44**(12), 1608–1622 (1998)
181. Kalashnikov, V.V., Maldonado, R.C.H., Camacho-Vallejo, J.-F., Kalashnykova, N.I.: A heuristic algorithm solving bilevel toll optimization problems. Int. J. Logist. Manag. **27**(1), 31–51 (2016)
182. Kalashnikov, V., Camacho, F., Askin, R., Kalashnykova, N.: Comparison of algorithms for solving a bi-level toll setting problem. Int. J. Innov. Comput., Inf. Control. **6**(8), 3529–3549 (2010)
183. González-Velarde, J.L., Camacho-Vallejo, J.-F., Pinto Serrano, G.: A scatter search algorithm for solving a bilevel optimization model for determining highway tolls. Computación y Sistemas **19**(1), 05–16 (2015)
184. Yin, Y.: Multiobjective bilevel optimization for transportation planning and management problems. J. Adv. Transp. **36**(1), 93–105 (2002)
185. Fan, W.: Optimal congestion pricing toll design for revenue maximization: comprehensive numerical results and implications. Can. J. Civ. Eng. **42**(8), 544–551 (2015)
186. Constantin, I., Florian, M.: Optimizing frequencies in a transit network: a nonlinear bi-level programming approach. Int. Trans. Oper. Res. **2**(2), 149–164 (1995)
187. Brotcorne, L., Labbé, M., Marcotte, P., Savard, G.: A bilevel model for toll optimization on a multicommodity transportation network. Transp. Sci. **35**(4), 345–358 (2001)
188. Amouzegar, M.A., Moshirvaziri, K.: Determining optimal pollution control policies: An application of bilevel programming. Eur. J. Oper. Res. **119**(1), 100–120 (1999)
189. Bostian, M., Whittaker, G., Sinha, A., Barnhart, B.: Incorporating data envelopment analysis solution methods into bilevel multi-objective optimization. In: 2015 IEEE Congress on Evolutionary Computation (CEC), pp. 1667–1674. IEEE (2015)
190. Bostian, M., Whittaker, G., Barnhart, B., Färe, R., Grosskopf, S.: Valuing water quality tradeoffs at different spatial scales: an integrated approach using bilevel optimization. Water Resour. Econ. **11**, 1–12 (2015)
191. Sinha, A., Malo, P., Anton, F., Deb, K.: Multi-objective Stackelberg game between a regulating authority and a mining company: a case study in environmental economics. In: 2013 IEEE Congress on Evolutionary Computation (CEC-2013), pp. 478–485. IEEE Press (2013)
192. Wood, K.: Deterministic network interdiction. Math. Comput. Model. **17**(2), 1–18 (1993)
193. Israeli, E., Wood, K.: Shortest-path network interdiction. Networks **40**(2), 97–111 (2002)
194. Church, R.L., Scaparra, M.P., Middleton, R.S.: Identifying critical infrastructure: the median and covering facility interdiction problems. Ann. Assoc. Am. Geogr. **94**(3), 491–502 (2004)
195. Lim, C., Smith, J.C.: Algorithms for discrete and continuous multicommodity flow network interdiction problems. IIE Trans. **39**(1), 15–26 (2007)
196. Scaparra, M.P., Church, R.L.: A bilevel mixed-integer program for critical infrastructure protection planning. Comput. Operat. Res. **35**(6), 1905–1923 (2008)
197. Akbari-Jafarabadi, M., Tavakkoli-Moghaddam, R., Mahmoodjanloo, M., Rahimi, Y.: A trilevel r-interdiction median model for a facility location problem under imminent attack. Comput. Ind. Eng. **114**, 151–165 (2017)
198. Ramamoorthy, P., Jayaswal, S., Sinha, A., Vidyarthi, N.: An exact method for trilevel hub location problem with interdiction. Eur. J. Oper. Res. **319**(3), 696–710 (2024)
199. Nigudkar, S., Sinha, A., Jayaswal, S.: Solving nonlinear max–min problems via piecewise-linear over-approximation: application to facility interdiction under congestion. Working paper, Indian Institute of Management Ahmedabad (2026)
200. Küçükaydin, H., Aras, N., Altınel, I.K.: Competitive facility location problem with attractiveness adjustment of the follower: a bilevel programming model and its solution. Eur. J. Oper. Res. **208**(3), 206–220 (2011)

201. Alekseeva E., Kochetova N., Kochetov Yu, and Plyasunov A. A hybrid memetic algorithm for the competitive p-median problem. IFAC Proc. Vol. **42**(4), 1533–1537 (2009)
202. Calvete, H.I., Galé, C., Iranzo, J.A.: An efficient evolutionary algorithm for the ring star problem. Eur. J. Oper. Res. **231**(1), 22–33 (2013)
203. Uno, T., Katagiri, H., Kato, K.: An evolutionary multi-agent based search method for stackelberg solutions of bilevel facility location problems. Int. J. Innov. Comput., Inf. Control. **4**(5), 1033–1042 (2008)
204. Sun, H., Gao, Z., Jianjun, W.: A bi-level programming model and solution algorithm for the location of logistics distribution centers. Appl. Math. Model. **32**(4), 610–616 (2008)
205. Maldonado-Pinto, S., Casas-Ramírez, M.-S., Camacho-Vallejo, J.-F.: Analyzing the performance of a hybrid heuristic for solving a bilevel location problem under different approaches to tackle the lower level. Math. Probl, Eng (2016)
206. Jin, Q., Feng, S.: Bi-level simulated annealing algorithm for facility location. Syst. Eng. **2**, 007 (2007)
207. Ramamoorthy, P., Jayaswal, S., Sinha, A., Vidyarthi, N.: Multiple allocation hub interdiction and protection problems: model formulations and solution approaches. Eur. J. Oper. Res. **270**(1), 230–245 (2018)
208. Reisi, M., Gabriel, S.A., Fahimnia, B.: Supply chain competition on shelf space and pricing for soft drinks: a bilevel optimization approach. Int. J. Prod. Econom. **211**, 237–250 (2019)
209. Dan, T., Marcotte, P.: Competitive facility location with selfish users and queues. Oper. Res. **67**(2), 479–497 (2019)
210. Dan, T., Lodi, A., Marcotte, P.: Joint location and pricing within a user-optimized environment. EURO J. Comput. Optim. **8**(1), 61–84 (2020)
211. Maric, M., Stanimirovic, Z., Milenkovic, N., DJenic, A.: Metaheuristic approaches to solving large-scale bilevel uncapacitated facility location problem with clients'preferences. Yugosl. J. Oper. Res. ISSN: 0354-0243 EISSN: 2334-6043 **25**(3) (2014)
212. Bansal, V., Jayaswal, S., Sinha, A.: Capacitated multiple allocation hub location problems under the risk of interdiction: model formulations and solution approaches. Ann. Oper. Res. **332**(1), 213–251 (2024)
213. Smith, W.R., Missen, R.W.: Chemical Reaction Equilibrium Analysis: Theory and Algorithms. John Wiley & Sons, New York (1982)
214. Raghunathan, A.U., Biegler, L.T.: Mathematical programs with equilibrium constraints (mpecs) in process engineering. Computers & Chemical Engineering **27**(10), 1381–1392 (2003)
215. Clark, P.A., Westerberg, A.W.: Bilevel programming for steady-state chemical process design-i. fundamentals and algorithms. Comput. Chem. Eng. **14**(1), 87–97 (1990)
216. Kocvara, M., Outrata, J.V.: On the solution of optimum design problems with variational inequalities. Recent Adv. Nonsmooth Opt. pp. 172–192 (1995)
217. Kočvara, M.: Topology optimization with displacement constraints: a bilevel programming approach. Struct. Optim. **14**(4), 256–263 (1997)
218. Herskovits, J., Leontiev, A., Dias, G., Santos, G.: Contact shape optimization: a bilevel programming approach. Struct. Multidiscip. Optim. **20**(3), 214–221 (2000)
219. Christiansen, S., Patriksson, M., Wynter, L.: Stochastic bilevel programming in structural optimization. Struct. Multidiscip. Optim. **21**(5), 361–371 (2001)
220. Bendsoe, M.P.: Optimization of Structural Topology, Shape, and Material, vol. 414. Springer (1995)
221. Guo, X., Zhang, W., Zhang, L.: Robust structural topology optimization considering boundary uncertainties. Comput. Methods Appl. Mech. Eng. **253**, 356–368 (2013)
222. Pan, Z., Gao, X., Kui, W.: First-order topology optimization via inexact finite element analysis. Comput. Aided Des. **157**, 103466 (2023)
223. Albrecht, S., Ramirez-Amaro, K., Ruiz-Ugalde, F., Weikersdorfer, D., Leibold, M., Ulbrich, M., Beetz, M.: Imitating human reaching motions using physically inspired optimization principles. In: 2011 11th IEEE-RAS International Conference on Humanoid Robots (Humanoids), pp. 602–607. IEEE (2011)

224. Johnson, M., Aghasadeghi, N., Bretl, T.: Inverse optimal control for deterministic continuous-time nonlinear systems. In: 2013 IEEE 52nd Annual Conference on Decision and Control (CDC), pp. 2906–2913. IEEE (2013)
225. Mombaur, K., Truong, A., Laumond, J.-P.: From human to humanoid locomotion1an inverse optimal control approach. Auton. Robot. **28**(3), 369–383 (2010)
226. Suryan, V., Sinha, A., Malo, P., Deb, K.: Handling inverse optimal control problems using evolutionary bilevel optimization. In: 2016 IEEE Congress on Evolutionary Computation (CEC-2016), pp. 1893–1900. IEEE Press (2016)
227. Zhang, G., Liu, J., Dillon, T.: Decentralized multi-objective bilevel decision making with fuzzy demands. Knowl.-Based Syst. **20**, 495–507 (2007)
228. Qing, X., Zhu, D., Li, S.: The supply chain optimal contract design under asymmetrical information. Syst. Eng.-Theo. Pract. **4**, 003 (2007)
229. Garen, J.E.: Executive compensation and principal-agent theory. J. Polit. Econ. 1175–1199 (1994)
230. Cecchini, M., Ecker, J., Kupferschmid, M., Leitch, R.: Solving nonlinear principal-agent problems using bilevel programming. Eur. J. Oper. Res. **230**(2), 364–373 (2013)
231. Motto, A.L., Arroyo, J.M., Galiana, F.D.: A mixed-integer lp procedure for the analysis of electric grid security under disruptive threat. IEEE Trans. Power Syst. **20**(3), 1357–1365 (2005)
232. Arroyo, J.M.: Bilevel programming applied to power system vulnerability analysis under multiple contingencies. IET Generat. Trans. Distrib. **4**(2), 178–190 (2010)
233. Garcés, L.P., Conejo, A.J., García-Bertrand, R., Romero, R.: A bilevel approach to transmission expansion planning within a market environment. IEEE Trans. Power Syst. **24**(3), 1513–1522 (2009)
234. Jin, S., Ryan, S.M.: Capacity expansion in the integrated supply network for an electricity market. IEEE Trans. Power Syst. **26**(4), 2275–2284 (2011)
235. Jenabi, M., Ghomi, S.M.T.F., Smeers, Y.: Bi-level game approaches for coordination of generation and transmission expansion planning within a market environment. IEEE Trans. Power Syst. **28**(3), 2639–2650 (2013)
236. Wogrin, S., Pineda, S., Tejada-Arango, D.A.: Applications of bilevel optimization in energy and electricity markets. In: Bilevel Optimization: Advances and Next Challenges, pp. 139–168. Springer (2020)
237. Grimm, V., Schewe, L., Schmidt, M., Zöttl, G.: A multilevel model of the european entry-exit gas market. Math. Methods Oper. Res. **89**(2), 223–255 (2019)
238. Böttger, T., Grimm, V., Kleinert, T., Schmidt, M.: The cost of decoupling trade and transport in the european entry-exit gas market with linear physics modeling. Eur. J. Oper. Res. **297**(3), 1095–1111 (2022)
239. Schewe, L., Schmidt, M., Thürauf, J.: Global optimization for the multilevel european gas market system with nonlinear flow models on trees. J. Global Optim. **82**(3), 627–653 (2022)
240. Sinha, A., Pujara, D., Singh, H.K.: Decomposition of difficulties in complex optimization problems using a bilevel approach. In: 2024 IEEE Congress on Evolutionary Computation (CEC), pp. 1–8. IEEE (2024)
241. Sinha, A., Pujara, D., Singh, H.K.: Bilevel optimization-based decomposition for solving single and multiobjective optimization problems. In: International Conference on Evolutionary Multi-Criterion Optimization, pp. 88–102. Springer (2025)
242. Sinha, A., Arora, S., Pujara, D.: Autoopt: a dataset and a unified framework for automating optimization problem solving. Accept. Neural Inf. Process, Syst (2025)
243. Zoph, B., Vasudevan, V., Shlens, J., Le, Q.: Automl for large scale image classification and object detection. Google AI Blog **2**, 2017 (2017)
244. Real, E., Aggarwal, A., Huang, Y., Le, Q.V.: Regularized evolution for image classifier architecture search. In: Proceedings of the AAAI Conference on Artificial Intelligence, vol. 33, pp. 4780–4789 (2019)
245. Zoph, B., Le, Q.V. Neural architecture search with reinforcement learning. In: Proceedings of the International Conference on Learning Representations (ICLR) (2017)

246. Chen, L.-C., Collins, M., Zhu, Y., Papandreou, G., Zoph, B., Schroff, F., Adam, H., Shlens, J.: Searching for efficient multi-scale architectures for dense image prediction. Adv. Neural Inf. Process. Syst., **31** (2018)
247. Elsken, T., Metzen, J.H., Hutter, F.: Neural architecture search: a survey. J. Mach. Learn. Res. **20**(55), 1–21 (2019)
248. Sinha, A., Pankaj, P.: A linear programming enhanced genetic algorithm for hyperparameter tuning in machine learning. In: 2023 IEEE Congress on Evolutionary Computation (CEC), pp. 1–8. IEEE (2023)
249. Liu, Y., Sun, Y., Xue, B., Zhang, M., Yen, G.G., Tan, K.C.: A survey on evolutionary neural architecture search. IEEE Trans. Neural Netw. Learn. Syst. **34**(2), 550–570 (2023)
250. Kaveh, M., Saadi Mesgari, M.: Application of meta-heuristic algorithms for training neural networks and deep learning architectures: a comprehensive review. Neural Process. Lett. **55**(4), 4519–4622 (2023)
251. Yao, Q., Wang, M., Chen, Y., Dai, W., Li, Y.-F., Tu, W.-W., Yang, Q., Yu, Y.: Taking human out of learning applications: a survey on automated machine learning. ArXiv preprint arXiv:1810.13306, 31 (2018)
252. Jaafra, Y., Laurent, J.L., Deruyver, A., Naceur, M.S.: Reinforcement learning for neural architecture search: a review. Image Vis. Comput. **89**, 57–66 (2019)
253. Parker-Holder, J., Rajan, R., Song, X., Biedenkapp, A., Miao, Y., Eimer, T., Baohe Zhang, V., Nguyen, R.C., Faust, A., et al.: Automated reinforcement learning (autorl): A survey and open problems. Journal of Artificial Intelligence Research **74**, 517–568 (2022)
254. Ogunsanya, M., Isichei, J., Desai, S.: Grid search hyperparameter tuning in additive manufacturing processes. Manufacturing Letters **35**, 1031–1042 (2023)
255. Fathima M Dhilsath and S Justin Samuel. Hyperparameter tuning of ensemble classifiers using grid search and random search for prediction of heart disease. *Computational Intelligence and Healthcare Informatics*, pages 139–158, 2021
256. Siti Fairuz Mat Radzi, Muhammad Khalis Abdul Karim, M Iqbal Saripan, Mohd Amiruddin Abd Rahman, Iza Nurzawani Che Isa, and Mohammad Johari Ibahim. Hyperparameter tuning and pipeline optimization via grid search method and tree-based automl in breast cancer prediction. *Journal of personalized medicine*, 11(10):978, 2021
257. Pier Paolo Ippolito. Hyperparameter tuning: the art of fine-tuning machine and deep learning models to improve metric results. In *Applied data science in tourism: Interdisciplinary approaches, methodologies, and applications*, pages 231–251. Springer, 2022
258. Leila Zahedi, Farid Ghareh Mohammadi, Shabnam Rezapour, Matthew W Ohland, and M Hadi Amini. Search algorithms for automated hyper-parameter tuning. arXiv preprint arXiv:2104.14677, 2021
259. Shekar BH and Guesh Dagnew. Grid search-based hyperparameter tuning and classification of microarray cancer data. In *2019 second international conference on advanced computational and communication paradigms (ICACCP)*, pages 1–8. IEEE, 2019
260. Bergstra, J., Bengio, Y.: Random search for hyper-parameter optimization. The journal of machine learning research **13**(1), 281–305 (2012)
261. James Bergstra, Rémi Bardenet, Yoshua Bengio, and Balázs Kégl. Algorithms for hyper-parameter optimization. *Advances in neural information processing systems*, 24, 2011
262. Eggensperger, K., Feurer, M., Hutter, F., Bergstra, J., Snoek, J., Hoos, H., Leyton-Brown, K.: Towards an empirical foundation for assessing bayesian optimization of hyperparameters. In NIPS workshop on Bayesian Optimization in Theory and Practice **10**, 1–5 (2013)
263. Snoek, J., Larochelle, H., Adams, R.P.: Practical Bayesian optimization of machine learning algorithms. Adv. Neural Inf. Process. Syst. **25** (2012)
264. Klein, A., Falkner, S., Bartels, S., Hennig, P., Hutter, F.: Fast Bayesian optimization of machine learning hyperparameters on large datasets. In: Artificial Intelligence and Statistics, pp. 528–536. PMLR, (2017)
265. Alibrahim, H., Ludwig, S.A.: Hyperparameter optimization: Comparing genetic algorithm against grid search and Bayesian optimization. In: 2021 IEEE Congress on Evolutionary Computation (CEC), pp. 1551–1559. IEEE (2021)

266. Lorenzo, P.R., Nalepa, J., Kawulok, M., Ramos, L.S., Pastor, J.R.: Particle swarm optimization for hyper-parameter selection in deep neural networks. In: Proceedings of the Genetic and Evolutionary Computation Conference, pp. 481–488 (2017)
267. Byla, E., Pang, W.: Deepswarm: optimising convolutional neural networks using swarm intelligence. In: UK Workshop on Computational Intelligence, pp. 119–130. Springer (2019)
268. Suganuma, M., Shirakawa, S., Nagao, T.: A genetic programming approach to designing convolutional neural network architectures. In: Proceedings of the Genetic and Evolutionary Computation Conference, pp. 497–504 (2017)
269. Chu, X., Zhang, B., Ma, H., Xu, R., Li, Q.: Fast, accurate and lightweight super-resolution with neural architecture search. In: 2020 25th International Conference on Pattern Recognition (ICPR), pp. 59–64. IEEE (2021)
270. Maziarz, K., Tan, M., Khorlin, A., Georgiev, M., Gesmundo, A.: Evolutionary-neural hybrid agents for architecture search. ArXiv preprint arXiv:1811.09828 (2018)
271. Guo, Yu., Li, J.-Y., Zhan, Z.-H.: Efficient hyperparameter optimization for convolution neural networks in deep learning: A distributed particle swarm optimization approach. Cybern. Syst. **52**(1), 36–57 (2020)
272. Yanyan Fan, Yu., Zhang, B.G., Luo, X., Peng, Q., Jin, Z.: A hybrid sparrow search algorithm of the hyperparameter optimization in deep learning. Mathematics **10**(16), 3019 (2022)
273. Zoph, B., Le, Q.V.: Neural architecture search with reinforcement learning. ArXiv preprint arXiv:1611.01578 (2016)
274. Chu, X., Zhang, B., Xu, R.: Multi-objective reinforced evolution in mobile neural architecture search. In: European Conference on computer vision, pp. 99–113. Springer (2020)
275. Mohan, A., Benjamins, C., Wienecke, K., Dockhorn, A., Lindauer, M.: Autorl hyperparameter landscapes. ArXiv preprint arXiv:2304.02396 (2023)
276. Hsu, C.-H., Chang, S.-H., Liang, J.-H., Chou, H.-P., Liu, C.-H., Chang, S.-C., Pan, J.-Y., Chen, Y.-T., Wei, W., Juan, D.-C.: Monas: multi-objective neural architecture search using reinforcement learning. ArXiv preprint arXiv:1806.10332 (2018)
277. Gao, Y., Yang, H., Zhang, P., Zhou, C., Hu, Y.: Graphnas: graph neural architecture search with reinforcement learning. ArXiv preprint arXiv:1904.09981 (2019)
278. Maclaurin, D., Duvenaud, D., Adams, R.: Gradient-based hyperparameter optimization through reversible learning. In: International Conference on Machine Learning, pp. 2113–2122. PMLR, (2015)
279. Sinha, A., Gunwal, S.: A linear programming-based hyper local search for tuning hyperparameters. Oper. Res. Lett. **61**, 107287 (2025)
280. Shi, W., Chang, Y., Gu, B.: Double momentum method for lower-level constrained bilevel optimization (2024)
281. Liu, H., Simonyan, K., Yang, Y.: DARTS: differentiable architecture search. ArXiv preprint arXiv:1806.09055 (2018)
282. Sinha, A., Khandait, T., Mohanty, R.: A gradient-based bilevel optimization approach for tuning regularization hyperparameters. Optim. Lett. **18**(6), 1383–1404 (2024)
283. Shi, W., Bin, G.: Improved penalty method via doubly stochastic gradients for bilevel hyperparameter optimization. In: Proceedings of the AAAI Conference on Artificial Intelligence vol. 35, pp. 9621–9629 (2021)
284. Franceschi, L., Frasconi, P., Salzo, S., Grazzi, R., & Pontil, M.: Bilevel programming for hyperparameter optimization and meta-learning. In: International Conference on Machine Learning, pp. 1568–1577. PMLR (2018)
285. MacKay, M., Vicol, P., Lorraine, J., Duvenaud, D., Gross, R.:. Self-tuning networks: Bilevel optimization of hyperparameters using structured best-response functions. ArXiv preprint arXiv:1903.03088 (2019)

A Review of Advances in Multi-objective Optimization for Facility Location Problems

Arup Kumar Bhattacharjee and Anirban Mukhopadhyay

Abstract Facility location problems play a vital role in transportation, healthcare, humanitarian logistics, and sustainable mobility, requiring a careful balance between efficiency, equity, and environmental sustainability. This article conducts a structured comparative review of ten representative studies to synthesize advancements in multi-objective optimization for facility location problems, rather than a full systematic literature review. The literature is categorized into five domains: hub location issues, emergency medical services, humanitarian and disaster logistics, electric vehicle infrastructure, and methodological advances in decision-support innovations, providing a focused synthesis of methodological and thematic progress in the field. A comparative analysis demonstrates a range of methodologies, including mathematical programming and metaheuristics to simulation-based and interactive optimization approaches, each addressing trade-offs among cost, accessibility, robustness, and fairness. Themes include the prevalence of multi-objective trade-offs, necessity for uncertainty management, integration of infrastructure and service provision, and increasing significance of equality considerations. Persistent research gaps such as limited scalability, lack of dynamic adaptation, insufficient stakeholder integration, incomplete sustainability metrics, and the need for hybrid optimization with machine learning are identified as key directions for future work. By bridging application-driven studies with methodological advances, this review provides an integrated perspective to guide both researchers and practitioners toward resilient, human-centered facility location strategies.

Keywords Multi-objective optimization · Facility location problem · Multi-objective facility location problems · Electric vehicles · Hub location problem · Metaheuristic approaches

---

A. K. Bhattacharjee (✉)
Department of Computer Science and Engineering, RCC Institute of Information Technology, Kolkata, India
e-mail: arupkumar.bhattacharjee@rcciit.org.in

A. Mukhopadhyay
Department of Computer Science and Engineering, University of Kalyani, Kalyani, India
e-mail: anirban@klyuniv.ac.in

S. Mukhopadhyay et al. (eds.), *Contemporary Advancement in Evolutionary Multi-objective Optimization*, Genetic and Evolutionary Computation,
https://doi.org/10.1007/978-981-95-7908-2_7

1 Introduction

Facility location problem (FLP) constitutes a core area within operations research, logistics, and systems engineering, with broad applications in transportation planning, emergency response, healthcare delivery, and supply chain management [9, 14]. The primary objective of FLPs are to identify optimal locations for facilities like hubs, medical centers, charging stations, or refugee camps, and thereafter allocate demand nodes to these facilities in an effective manner [1, 7]. In reality, these kind of selections are very complex in nature since they require balancing of multiple conflicting goals, like lowering costs, making services easier to get, lowering environmental impacts, and making sure everyone has equal access to services [3, 4, 6]. Since these problems are quite complicated and have so many goals, so they need advanced optimization models and ways to solve them. It is important to clarify that the present work does not follow the formal protocol of a Systematic Literature Review. Instead, this article has adopted a structured analytical approach, where ten representative and high-impact studies were purposively selected based on their methodological diversity, application relevance, and citation significance. This focused comparative framework allows for in-depth examination of multi-objective facility location problems across FLP domain while maintaining analytical rigor and thematic coherence.

Recent research shows that multi-objective facility location problem (MOFLP) can be encountered in a wide range of situations [1]. Many subcategories of FLP like hub location problem (HLP), have evolved from fundamental cost minimization models to complex multi-objective frameworks that incorporate mobility, depreciation, and environmental considerations. Trade-offs between costs and travel time between hubs and demand nodes, incorporating dynamic demand and environmental concerns into hub location problems, life-saving actions in the fields of health care and emergency response, disaster management, and humanitarian work to address refugee camp location challenges or incorporating migration simulations are some of the popular areas under MOFLPs [15]. The growth of electric vehicles (EVs) and green logistics has made location problems even harder to solve.

Even though these problems are targeted by many researchers, still there are many intrinsic problems that exist. Most models are tested on small datasets, so we don't know if they will work in larger or real-time situations. Demand uncertainty, disasters, and infrastructure conditions are often oversimplified, but real-world applications need strong ways to adapt [2, 17]. Additionally, sustainability aspects like resilience, life-cycle effects, and climate adaptation were insufficiently represented in research domain. The literature review shows both how far the field has come and how much more work needs to be done. It shows that we need approaches that are integrated, scalable, and stakeholder-centric.

This article systematically evaluates ten representative papers covering hub location, emergency medical services and healthcare, humanitarian logistics, EV infrastructure, and methodological innovations. This review segregates the entire study into four parts: (i) classification schemes (Sect. 3), (ii) comparative methodological frameworks (Sect. 4), (iii) cross-cutting themes (Sect. 5), and (iv) research gaps and future directions (Sect. 6). This review aims to provide a unified viewpoint on the influence of multi-objective optimization on modern strategies for FLP by connecting problem-specific research with methodological advancements.

2 Background and Motivation

Detailed investigation of FLP constitutes a fundamental aspect of operations research, originated from classical location theory and transportation science. Early models were usually single-objective, which meant they tried to minimize costs or distances. In last two decades, on the other hand, have seen a clear shift toward multi-objective frameworks. This is because modern systems are becoming more and more complex. This change is not just theoretical, but they are driven by real-world needs in a variety of application areas. In this review article we have mainly focused on the following areas:

1. **Urbanization and Infrastructure Stress**: Rapid urbanization has emphasized the need for better transportation and logistics systems, but it has also made land use, traffic, and environmental sustainability more difficult. Hub-spoke network, distribution centers, and transport corridors need to be built in a way that balances cost-effectiveness with environmental and social concerns. For example, Mokhtarzadeh et al. and Pourghader Chobar et al. conducted studies focused on the optimal placement of hubs, explicitly considering environmental factors such as noise and emissions during the optimization process. [16, 18]. This type of integration highlights the significance of mitigating the adverse impacts of urban infrastructure while addressing the increasing demand.
2. **Climate Change and Disaster Preparedness**: The growing number and severity of climate-related disasters, like earthquakes, floods, and storms, have made it even more important to plan for resilient facility locations. Studies on humanitarian logistics and emergency response underscore the necessity of multi-objective trade-offs among cost, speed, uncertainty, and rapid change [13, 20]. The strategic placement of emergency facilities can determine the difference between prompt assistance and disastrous delays.This scenario highlights the critical need for strong, flexible, and multi-objective optimization models to protect human lives.
3. **Healthcare Accessibility and Equity**: Healthcare systems all over the world have to deal with two big problems, not enough resources and unfair access, especially in rural or underprivileged areas. Research on facility location in this area shows that coverage, fairness, and response times are just as important as efficiency. Chen and Lai show that building or upgrading rural EMS facilities can significantly reduce the population lacking health insurance coverage [8]. Khalilzadeh and Bahari also used fairness-based models to make sure that services are distributed fairly [12]. These contributions reflect broader policy concerns regarding universal health coverage and the ethical dimensions of infrastructure planning.

4. **Transition to Sustainable Mobility**: EVs are on high demand because of the global push for de-carbonization and sustainable transportation systems. Due to the limited driving ranges and charging needs, EVs have different operational limits than regular cars. This has led to electric location-routing problems (ELRP), where decisions about routing and placement of their service centers are closely linked. Wang et al. address this issue by suggesting accelerated evolutionary algorithms to enhance EV logistics [19]. These kinds of problems show how new technological advances add new levels of complexity that need new ways to optimize.
5. **Decision Support and Stakeholder Involvement**: Development of optimization techniques shows that more and more people realize that decision-makers not only find mathematically optimal solutions but also understand and use them correctly. Conventional metaheuristics frequently produce extensive collections of pareto-optimal solutions, which may overburden users. The graph-theoretical clustering method by Kahagalage et al. and the interactive evolutionary framework by Barbati et al., both are important approaches toward making optimization outputs more in line with user decisions [5, 11]. These methods consider preference input, solution pruning, and interpretability for guaranteeing that advanced algorithms apply significant and definite influence in practical applications.

2.1 Motivation for This Review

The coming together of these global issues, namely, urbanization, climate change, healthcare equity, sustainable mobility, and decision support makes research on where and how to use/ place facilities more important than ever. While individual studies concentrate on particular applications, there is an increasing demand for comprehensive reviews that amalgamate methodological innovations, pinpoint shared challenges, and emphasize prospects for interdisciplinary learning. This article seeks to furnish an integrated perspective by classifying, comparing, and critically analyzing the ten representative studies examined herein, yielding insights that are pertinent both to particular problem areas and to broader contexts.

2.2 Global Challenges and Their Technical Implications for Multi-objective Optimization Approaches

The global concerns outlined in Sect. 2.1 such as rapid urbanization, climate change, healthcare inequity, and the transition to sustainable mobility—directly translate into complex technical and research challenges for Multi-objective Optimization (MOO) domain.

1. Urbanization and Infrastructure Stress: High population density and dynamic urban growth demand scalable and adaptive optimization models capable of handling large, evolving networks. This necessitates Multi-objective Optimization Approaches (MOOA) techniques that integrate real-time data and dynamic demand forecasting.
2. Climate Change and Disaster Preparedness: Increasing frequency of extreme events introduces uncertainty and nonstationarity in system parameters. Robust and stochastic MOOA models are required to ensure resilience under variable conditions.
3. Healthcare Accessibility and Equity: Addressing social fairness in facility location leads to multi-objective formulations that balance efficiency with accessibility and equality. This creates new modeling challenges in defining and quantifying fairness metrics within optimization frameworks.
4. Sustainable Mobility Transition: Electrification and de-carbonization introduce coupling between infrastructure and service networks (e.g., charging location and route optimization). These problems demand hybrid MOOA formulations combining discrete and continuous decision variables.
5. Decision-Making and Stakeholder Integration: The complexity of real-world facility location decisions requires interactive and preference-driven MOOA frameworks that can incorporate human judgment, multi-agent objectives, and conflicting stakeholder priorities.

By relating these global issues to the technical and methodological frontiers of MOOA, this section underscores the interdisciplinary nature of facility location research and the need for algorithms that are both computationally efficient and socially responsive.

3 Classification of Facility Location Problems and Application Domains

The literature examined regarding MOFLPs can be classified into various categories, emphasizing the variety of problem domains, application contexts, and methodological approaches. This classification is crucial as it illustrates that facility location issues are not homogeneous rather, they vary based on the operational context, decision-making objectives, and stakeholder interests.

Hub location problems (HLPs) maybe the most common type of FLP. These problems are integral to transportation and logistics systems, where hubs serve as pivotal sites for the aggregation and distribution of commodities, information, or passengers. In the past, HLP mainly focused on reducing travel costs or time. Recent research, on the other hand, has significantly expanded the scope to encompass environmental and social issues. Mokhtarzadeh et al. formulated a p-mobile hub location problem that acknowledges the requirement for periodic hub relocation, encompassing depreciation and relocation costs [16]. This paper adds a time and sustainability aspect that earlier static models didn't consider. Demir et al. addressed the capacitated multiple allocation hub location problem, necessitating the equilibrium of assignment and routing with limited hub capacity [10]. Pourghader Chobar et al. proposed a model that incorporates dynamic demand variations and environmental goals, thereby reconciling hub location models with sustainability requirements [18]. These three papers together demonstrate the evolution of HLP research into a multifaceted discipline that addresses efficiency, sustainability, and adaptation at the same time.

The second classification type includes problem domain involving healthcare and EMS facilities. Since hub location problems focus on making things work better and more efficiently, similarly emergency medical services (EMS) under FLP purview prioritize the facilitation of healthcare services. Chen and Lai developed models for rural China that considered the improvement of existing road infrastructure, the establishment of new EMS facilities, or a combination of both approaches [8]. Their research is significant for demonstrating how assumptions regarding infrastructure, such as the feasibility of constructing new roads, significantly influence optimal solutions, and highlights the relationship between facility location and supporting infrastructure. Khalilzadeh and Bahari developed a fairness-driven methodology utilizing the P-envy framework, ensuring that no community faces disproportionately inferior service access relative to others [12]. Their model implements fairness using a mathematical programming framework, illustrating that equity can be integrated into optimization. These studies constitute a crucial research domain that prioritizes human lives and social justice within the field of location science.

The third group is logistics for humanitarian and disaster response. These issues are different from EMS because they deal with sudden crises that affect people who have been displaced, and infrastructure that has been damaged. Xue et al. developed a multi-objective simulation optimization model for the establishment of refugee camps, explicitly integrating population migration into the optimization procedure [20]. Their model includes factors like environmental suitability and survival conditions in addition to cost, coverage, and accessibility. Khodadadi et al. put forward a multi-objective model for earthquake response that includes facility location, routing, and resource allocation [13]. This model covers not only where to place facilities but also how to provide emergency supplies to victims and to distribute resources to them. Their strong, multi-period framework shows how disaster logistics really work over time and in random ways. This category stresses that when facility location is used in crisis situations, it becomes a matter of human survival, which calls for strong and flexible solution.

A fourth type of classification under consideration is the infrastructure for EVs and sustainable mobility. Wang et al. examined the ELRP, which involves the simultaneous optimization of the placement of EV charging stations and routing choices [19]. In traditional logistics, problems with facilities and routing were often solved separately. However, due to range limits and charging dependencies, EV systems need to be tightly integrated. This new problem is a result of society's need for de-carbonization and long-lasting infrastructure.

Finally, the fifth category includes new ways to help people make decisions. These studies concentrate less on a specific domain and more on advancement of the tools accessible to decision-makers. Kahagalage et al. proposed a graph-theory-based clustering methodology to diminish the magnitude of pareto-optimal solution sets, thereby assisting decision-makers in traversing intricate trade-off domains [11]. Barbati et al. developed an interactive evolutionary algorithm (NEMO-II-Ch) that integrates decision-maker preferences into the optimization process, which helps determine the solutions [5]. These methodological contributions are essential as they bridge the gap between algorithmic output and practical decision-making.

In this way, the classification scheme shows five thematic approach, namely, problems with finding a hub location, finding a location for EMS and healthcare facilities, humanitarian logistics, EV infrastructure, and new tools and methods for decision-making. Each approach represents unique application requirements and solution approaches, collectively creating a cohesive understanding of how multi-objective optimization enhances research under FLP.

4 Comparative Methodology and Technical Challenges in Multi-objective Optimization for Facility Location Problems

The methodological landscape of FLP research is as diverse as its applications. A comparison of the studies shows that there is a relationship between exact mathematical programming methods, metaheuristic algorithms, simulation-based optimization, and decision-support improvements. Each methodology has its own pros and cons, and which one to use depends on the size, complexity, and context of the decision-making process.

Mathematical programming formulations, such as mixed-integer linear programming (MILP) and mixed-integer nonlinear programming (MINLP), have traditionally been used to solve classical FLPs. Mokhtarzadeh et al. conceptualized the p-mobile hub locationâŁ"allocation problem as a MINLP, achieving optimal solutions for small and medium-sized instances [16]. In the same way, Khodadadi et al. used a strong multi-objective MILP for representing use of logistics after earthquake and have shown how formal mathematical structures can capture multiple goals, keeping constraints logically consistent [13]. Khalilzadeh and Bahari employed mathematical programming to implement their P-envy fairness concept, demonstrating the adaptability of traditional formulations to incorporate equity considerations [12]. The power of mathematical programming lies in its precision and ability to identify exact solutions, but it doesn't scale well specially when dealing with large or nonlinear problems.

To get around these problems, most recent works use metaheuristic algorithms. Some of these are genetic algorithms (GA), simulated annealing (SA), particle swarm optimization (PSO), non-dominated sorting genetic algorithm II (NSGA-II), and multi-objective particle swarm optimization (MOPSO). Demir et al. executed a comprehensive comparison of various metaheuristics, utilizing them for the capacitated hub location problem and evaluating both solution quality and computational efficiency [10]. In Demir et al., the mathematical model of the capacitated multiple allocation hub location problem is built around two objective functions and a set of constraints, expressed through Eqs. (1)–(8) [10]. Equation (1) defines the primary objective, which seeks to minimize the total system cost. Total cost includes transportation costs along arcs, fixed establishment costs for hubs, and operating costs associated with activated arcs. Equation (2) introduces a second objective that minimizes the maximum travel time experienced by any commodity across the network. Equations (3)–(8) are the constraints associated with the objective functions. Equation (3) specifies that at least one hub must be established, but the number of hubs cannot exceed the total number of nodes in the network. Equation (4) enforces that the total flow traversing any arc cannot exceed its designated capacity, thereby preserving arc feasibility. Similarly, Eq. (5) ensures that the cumulative inbound flow handled by any hub does not surpass its hub capacity. Equation (6) requires that the total demand of each commodity is satisfied by summing the demands of its disaggregated packages, guaranteeing demand satisfaction. Equation (7) links the routing variables with hub decisions, ensuring that flows can only be routed through nodes that are established as hubs. Finally, Eq. (8) adds bounds on commodity flows over arcs, such that the flow of each commodity on an arc cannot exceed the total flow capacity already defined for that arc. Together, these equations integrate cost minimization, service quality, and capacity management into a complex multi-objective framework, while the constraints ensure the logical and operational validity of the solution. Tables 1 and 2 give brief description of the decision variables and parameters used in these equations.

$$\min Z_1 = \sum_a F_a C_a \left(\chi x'_{au} x_{av} + \alpha x_{au} x_{av} + \delta x_{au} x'_{av}\right) + \sum_a \lambda_a Z_a + \sum_{i=1}^{N} K_i x_i \quad (1)$$

$$\min Z_2 = \max_{k,p} \sum_a T_a A^a_{kp} \quad (2)$$

Subject to the following constraints.

$$1 \leq \sum_i x_i \leq N, \quad (3)$$

$$F_a \leq Q_a \quad \forall a \in E, \quad (4)$$

$$\sum_{a \in V^+(h)} F_a \leq Q_h \quad \forall h \in H, \quad (5)$$

Table 1 Description of the decision variables used in Eqs. (1) and (2)

Decision variable	Type of value	Description
x_i	Binary	$x_i = 1$, if node i is selected as a hub, 0 otherwise
x_i'	Binary	$x_i' = 1$, if node i is selected as a spoke, 0 otherwise
Z_a	Binary	$Z_a = 1$, if arc a is used in the network, 0 otherwise
A_{kp}^a	Binary	$A_{kp}^a = 1$, if package p of commodity k uses arc a, 0 otherwise
f_a^k	Continuous	Flow of commodity k over arc a

Table 2 Description of the parameters mentioned in Eqs. (1) and (8)

Parameters	Description	Parameters	Description
a_u	Initial nodes of arc a belongs to E	a_v	Terminal nodes of arc a belongs to E
F_a	Total flow traversing arc a	C_a	Unit transportation cost on arc a
T_a	Travel time over arc a	λ_a	Fixed cost for establishing arc a
Q_a	Capacity of arc a	Q_h	Capacity of hub h
K_i	Fixed cost of establishing hub h	i_k	Source nodes for commodity k
j_k	Destination nodes for commodity k	D_k	Total demand of commodity k
D_{kp}	Demand of package p of commodity k	χ	Discount factors for collection
α	Discount factors for transfer	δ	Discount factors for distribution

$$\sum_p D_{kp} = D_k \quad \forall k, \tag{6}$$

$$A_{kp}^a \leq x_{av} Q_h, \quad a_v \neq j_k, \ \forall k, \tag{7}$$

$$0 \leq f_a^k \leq F_a \quad \forall a \in E. \tag{8}$$

By formulating these mathematical models, Demir et al. applied various metaheuristic algorithms to the bi-objective capacitated hub location problem. Similarly, Pourghader Chobar et al. have also used hybrid metaheuristics to solve the multi-objective hub problem with changing demand [18]. Metaheuristics are great at searching through large solution spaces and making good guesses about the Pareto frontier. The limitation of these approaches is that they don't always find the best solution, rather optimal solution, and their performance is often dependent on the problem, i.e., the operators and parameters need to be chosen carefully.

Simulation-based optimization is another approach for solving MOFLPs. Simulation models are combined with evolutionary algorithms to test solutions in situations that are complex or random. Xue et al. illustrate their study on refugee camp locations, integrating agent-based migration simulations with evolutionary optimization to address multi-objective challenges [20]. This method is very useful for problems where it's hard to figure out the nature of the solution, but it can take a lot of computing power and planning to avoid biases in the simulation.

Finally, we have explored few new methods, which are trying to make decision-support systems easier to use. Kahagalage et al. (2023) claim that generating a pareto frontier is inadequate if decision-makers are unable to evaluate numerous non-dominated solutions [11]. Their graph-theoretical clustering method cuts down on the number of solutions without losing diversity, which makes it easier to take decisions. They addressed the challenge of large Pareto fronts by introducing a graph-theoretical clustering approach. They employed the Silhouette index to evaluate clustering quality (Eq. (9)). In Eq. (9), $a(i)$ represents intra-cluster distance for cluster (of nodes) i and $b(i)$ represents mean nearest-cluster distance for cluster i. Each $s(i)$ determines the compactness of the cluster i and their summation determines the overall quality of the clustering.

$$s(i) = \frac{b(i) - a(i)}{\max\{a(i), b(i)\}} \tag{9}$$

Barbati et al. (2024) enhance this concept by integrating decision-maker preferences into the optimization process, resulting in solutions that are both mathematically optimal and congruent with human priorities [5]. These contributions advance the field toward "decision-centric optimization," emphasizing not only computational performance but also the integration of optimization with actionable insight. The comparative methodological framework shows a range of approaches, from exact methods that are mathematically sound but hard to compute, to metaheuristics that are flexible but based on some rules of thumb. The future of facility location research probably lies in hybrid methods that combine these strengths: heuristics for large-scale exploration, mathematical rigor for smaller problem components, simulation for modeling uncertainty, and decision-support tools for practical usability.

4.1 Technical Challenges in Multi-objective Facility Location Problems

In this section we have compared the methodologies used for MOFLPs, and implementation of MOOA introduces a distinct set of technical challenges such as:

1. Genotype Formation: The choice of representation significantly affects search performance and solution quality. FLPs often employ binary encoding (facility open/close decisions), integer encoding (assignment and capacity levels), or hybrid and permutation-based structures (for routing integrated FLPs). Designing an efficient genotype requires balancing exploration capability with problem-specific constraints, such as capacity limits, demand coverage, and geographical restrictions.
2. Choice and Design of Objectives: FLPs are inherently multidimensional, and the definition of objective functions determine the quality and interpretability of results. Common objectives include minimization of total cost, travel time, or emissions, and maximization of coverage, equity, or service quality.

Based on these methodological frameworks, we may infer that future researchers may focus on:

1. Adaptive or self-evolving genotype encoding that adjusts to dynamic problem scales.
2. Multi-level objective structures incorporating both operational and strategic goals.
3. Hybrid MOOA frameworks integrating data-driven prediction with optimization.
4. Interpretable optimization outputs aligned with stakeholder preferences.

Addressing these technical challenges may enhance the robustness, scalability, and practical relevance of MOOA-based facility location models.

5 Cross-cutting Themes

Even though the papers that were looked at cover a lot of different application areas, there are a few common themes that run through all of them and show that there are common concerns across domains. The primary and most significant theme is the multi-objective characteristic of FLPs. All studies under consideration have more than one goal, and these goals often conflict with each other. For example, in hub systems, the goal is to keep costs low, in EMS, the goal is to be fair, and in refugee camps, the goal is to find a place that is both safe and good for the environment. This signifies a comprehensive transition in optimization research from singular efficiency to multidimensional trade-offs that encapsulate real-world complexities. A second frequently occurring theme is uncertainty and the mechanism to deal with uncertainty in FLP. Numerous applications, especially in disaster logistics and dynamic demand hubs, entail uncertain conditions that cannot be adequately represented by deterministic assumptions. Khodadadi et al. utilize robust optimization to address variability in disaster scenarios [13]. Mokhtarzadeh et al. and Pourghader Chobar et al. incorporate dynamic and stochastic demand patterns, guaranteeing that solutions remain viable amidst variations [16, 18]. This shows that people understand that choosing a location for a facility is not just a one-time optimization problem, but a process that changes over time and is based on uncertainty.

The third theme is the connection between infrastructure and service delivery. In EMS (Chen and Lai) and ELRP (Wang et al.), facility location is inseparably linked to operational factors, including road quality, response times, and routing feasibility [8, 19]. This integration represents real-world scenario, where the placement of infrastructure and the delivery of services must be optimized together rather than separately. A fourth theme is the conflict between fairness and efficiency. In logistics and hub models, goals that are based on efficiency are the most important. In healthcare and humanitarian settings, on the other hand, goals that are based on fairness are becoming more important. Khalilzadeh and Bahari explicitly incorporate fairness into EMS allocation, whereas Xue et al. reconcile environmental suitability with refugee welfare [12, 20]. This theme emphasizes the ethical aspect of facility placement, especially in situations where choices impact vulnerable groups.

Finally, a strong recently used methodological theme is the innovative application of decision-support systems. Kahagalage et al. and Barbati et al. illustrate a movement toward enhancing the interpretability and applicability of optimization outputs [5, 11]. This indicates that people recognize optimization as a component of a broader decision-making framework that encompasses human judgment and institutional priorities. These cross-cutting themes show that facility location research has the same methodological and ethical issues, even though the domains are different. The convergence of multi-objectivity, uncertainty management, infrastructure and service integration, equity, and decision-support innovation indicates that forthcoming facility location models will be progressively comprehensive, resilient, and human-centric.

Brief summary of Sects. 3, 4 and 5 can be found in Table 3 for the papers under consideration.

6 Research Gaps and Future Directions

Even though the literature review shows that progress has been made, there are still some gaps in FLP research. Filling in these gaps is a way to move the domain of MOFLP forward.

One big area of concern is that it doesn't work in the real-world scenario and isn't scalable to large problem domain. Numerous studies depend on benchmark datasets (e.g., CAB, Turkish or Australian Post) or regional case studies with restricted scope. These datasets are useful, but they hardly represent complicated global, real-time logistics networks. Future research needs to test models on big, diverse datasets, and it might be helpful to include live data streams from sensors, IoT, and geospatial systems.

The second gap is the combination of dynamic and real-time adaptation. Some models do take into account changing demand (Pourghader Chobar et al.) or disasters (Khodadadi et al.), but very few include real-time updates on traffic conditions, or how people are moving after disaster [13, 18]. The rise of real-time data sources

Table 3 Summary of classification scheme, comparative methodology[1] and themes of the papers

Paper description		Mokhtarzadeh et al. (2021)	Demir et al. (2022)	Pourghader chobar et al. (2021)	Chen and Lai (2022)	Khalilzadeh and Bahari (2023)	Xue et al. (2024)	Khodadadi et al. (2023)	Wang et al. (2023)	Kahagalage et al. (2023)	Barbati et al. (2024)
Classification scheme	Hub location problems	Yes	Yes	Yes							
	EMS facility location				Yes	Yes					
	Humanitarian/disaster response						Yes	Yes			
	EV and sustainable logistics								Yes		
	Methodological advances									Yes	Yes
Comparative methodology	Mathematical programming	Yes[2]			Yes	Yes		Yes	Yes		
	Metaheuristics		Yes	Yes							
	Simulation-based optimization						Yes				
	Decision-support enhancements									Yes	Yes
Cross-cutting themes	Multi-objective optimization	Yes	Yes	Yes	Yes	Yes	Yes	Yes	Yes	Yes	Yes
	Uncertainty and robustness	Yes		Yes			Yes	Yes			
	InfrastructureâŁ“service integration				Yes		Yes		Yes		
	Fairness and accessibility				Yes	Yes					
	Decision-support innovation		Yes							Yes	Yes

[1]Several studies use hybrid or combined frameworks, where mathematical models are complemented by metaheuristics or simulation-based optimization
[2]Mokhtarzadeh et al. (2021) used hybrid approach (k-Medoids clustering with a multi-objective particle swarm optimization)

presents an opportunity to create adaptive optimization systems those dynamically determine how facilities make decisions.

Another area of concern is about modeling that focuses on stakeholders. Barbati et al. present interactive preference input, however, the majority of studies operate under the assumption of a singular decision-maker or a unified set of objectives [5]. In reality, facility location often involves many different groups of people, such as government agencies, NGOs, and local communities, whose goals may not always be the same. Future research ought to investigate multi-agent decision-making frameworks, negotiation models, and participatory optimization to harmonize conflicting stakeholder interests.

A fourth gap exists in the metrics for sustainability and resilience. Broader sustainability issues like life-cycle emissions, climate resilience, and long-term adaptability are not being dealt with as consistently as environmental and social goals. Future research should broaden the objectives to include resilience to climate shocks, intergenerational equity, and systemic environmental effects.

Lastly, there is room for combining algorithms and machine and deep learning approaches. Most of the time, current methods use classical metaheuristics and mathematical programming. But new developments in reinforcement learning, surrogate modeling, and hybrid metaheuristic machine learning frameworks make it possible to find solutions faster, make them more scalable, and adaptable to changes in real time. For example, machine learning can find areas of the solution space that are likely to work well, and then metaheuristics can use those areas to refine the solution.

The future of facility location research lies in developing models for global, real-time contexts, incorporating dynamic adaptation, stakeholder perspectives, broader sustainability metrics, and hybrid algorithms. Addressing these challenges will enhance computational efficiency while keeping research relevant to global issues such as climate change, urbanization, healthcare equity, and sustainable mobility. Table 4 highlights the research gaps and future directions across domains.

7 Conclusion

The literature on MOFLP illustrates the variety of real-world challenges and the advancement of optimization methodologies designed to tackle them. This review has shown how facility location research has shifted its paradigm from models that only looked at cost minimization to multi-objective frameworks that take into account environmental, social, and equity issues. We have segregated our survey into five thematic groups: transportation hub systems, healthcare and EMS, humanitarian and disaster logistics, sustainable EV infrastructure, and methodological innovations. Each domain has its own set of important features to think about. HLP focus on adaptability and sustainability, EMS focuses on accessibility and fairness, humanitarian logistics focus on being strong in the face of uncertainty, EV infrastructure focuses on technological limits and sustainability goals, and finally methodological

Table 4 Research gap and future directions in the domain of MOFLP

Domain	Research gap	Future direction
Real-time and adaptive optimization	A major shortcoming identified across the reviewed literature is the lack of real-time responsiveness within current optimization frameworks. Most optimization models are static or based on historical and forecasted data. While these approaches provide valuable baseline insights, they fall short in dynamic environments where sudden shifts in demand, unforeseen emergencies, or infrastructure disruptions require immediate adaptation. Static systems hamper effectiveness in high-stakes urban applications such as traffic control, waste management, or emergency logistics	Advancment toward real-time, adaptive optimization systems should represent a key frontier for research. Future models should incorporate live data streams from sensors, vehicle telemetry, mobile devices, and user-generated feedback to enable continuous decision-making under uncertainty. The integration of Internet of Things (IoT) devices and edge computing architectures could allow for decentralized urban systems to autonomously respond to evolving conditions. Such frameworks would be particularly transformative in domains like traffic management, where real-time responsiveness can reduce congestion and enhance safety, and in emergency response, where timely reallocation of resources can save lives
Multi-system interoperability	Although several studies hint at the interdependencies among urban subsystems like EV charging and power grids, most models treat each infrastructure component in isolation. This approach overlooks the complex interactions and potential synergies between infrastructures, leading to suboptimal outcomes and inefficiencies. Independent planning of transportation and energy systems can result in grid overloads or underutilized charging assets, rather than leveraging coordinated strategies for load balancing or coexistence of related infrastructural components	There is an important need to develop integrated optimization frameworks that address multiple urban infrastructural domains such as energy, logistics, and waste management within a unified model. These frameworks should capture the technical interdependencies like shared energy flows, interoperable communication standards, and data exchange mechanisms. Effective implementation will require institutional collaboration across goverment offices, regulatory bodies, and private-sector stakeholders. Such interconnected systems ensure more efficient resource allocation, enhanced system resilience, and more sustainable urban development overall
Equity, accessibility, and social justice	Urban infrastructure planning must evolve beyond traditional metrics like cost-efficiency and environmental sustainability to address issues of equity and accessibility. Only a few reviewed studies like Smart Dustbins, 2024 have considered socio-demographic variability or aim to ensure fair distribution of services. As a result, many optimization models risk reinforcing existing disparities, benefiting affluent people while neglecting marginalized communities	To promote inclusive urban development, equity metrics should be embedded directly into the core of optimization frameworks. This must involve integrating equity metrics into objective functions, like prioritizing infrastructure development in underserved neighborhoods, or addressing issues faced by low-income groups, the elderly, or individuals with disabilities. Doing so will help ensure that smart infrastructure systems serve not just efficiently and sustainably, but also fairly

(continued)

Table 4 (continued)

Domain	Research gap	Future direction
Data quality and privacy concerns	Many urban infrastructure optimization models depend on high-resolution data related to traffic patterns, user behavior, and energy consumption. While such data can significantly enhance model precision, but many a times these datasets are incompleteness, inconsistency, and limited access. These data limitations not only compromise the accuracy and reliability of model outputs but also raise ethical and legal issues that can hinder public trust and adoption	To develop robust optimization techniques that perform well under data uncertainty and incorporates privacy-preserving computation approaches, such as federated learning, homomorphic encryption, and differential privacy, that enable models to learn from decentralized or sensitive data without compromising individual privacy. These innovations will allow urban optimization frameworks to function effectively in real-world settings while maintaining strong ethical standards, thereby promoting both technical performance and societal acceptance
Integration of artificial intelligence	Although simulation and forecasting techniques are used in several infrastructure-related studies, the application of advanced AI techniques, particularly deep learning and reinforcement learning, remains largely under utilized in infrastructure optimization. These AI approaches hold significant potential, as they are well-suited to capturing complex, nonlinear interactions and can continuously learn from dynamic urban environments over time	A promising avenue for future exploration lies in the development of hybrid models that integrate traditional optimization methods with cutting-edge AI techniques. Such models could offer enhanced adaptability and predictive accuracy, making them more responsive to the evolving nature of urban systems. For example, reinforcement learning could be utilized to identify optimal infrastructure policies in evolving urban contexts, thereby enabling more resilient and intelligent infrastructure planning
Scalability and computational efficiency	Many state-of-the-art optimization models—particularly those based on Mixed-Integer Linear Programming (MILP), Mixed-Integer Nonlinear Programming (MINLP), or bilevel programming offer high precision. However, these methods are often computationally demanding and face significant challenges when applied to large-scale problems, such as those involving entire metropolitan regions or nationwide infrastructure systems. Their complexity can limit practical deployment, especially when real-time or near-real-time decision-making is required	To overcome these scalability barriers, future work should prioritize the development of more efficient and scalable solution strategies. This may involve the use of decomposition techniques (e.g., Benders or Dantzig-Wolfe decomposition) to break down large problems into smaller, more manageable subproblems. Additionally, use of cloud-based optimization platforms can dramatically reduce computation time and enable the application of high-fidelity models to complex, city-wide or multi-city infrastructure planning

studies focus on making sure that solutions are easy to understand and use as well as time and space efficient.

After exploring different approaches to solve MOFLPs, we have found a range of tools that are available. Classical mathematical programming offers precision and rigor for small to medium-sized problems, whereas metaheuristics excel in large-scale, NP-hard scenarios by effectively approximating pareto frontiers. Simulation-based optimization fills the gap where analytical evaluation isn't possible, especially when things are uncertain or changing. More recently, decision-support methods like clustering and interactive optimization have made it possible for practitioners to use complex solution sets. These methods highlight the importance of employing a variety of approaches, i.e., no singular approach is adequate for all contexts, and the most effective models frequently integrate components of exact, heuristic, simulation, and decision-support methodologies.

Cross-cutting themes make the field even more coherent. The prevalence of multi-objectivity signifies an increasing recognition that efficiency alone is inadequate, decision-makers must reconcile economic, social, and environmental considerations. Integration between infrastructure and service delivery is a recurring necessity, whether simulteneous optimization of EMS roads and stations or linking EV routing with charging facilities. Equity becomes an ethical necessity, especially in healthcare and humanitarian contexts, while methodological progress highlights the importance of integrated perspectives and future directions in the review of advancements in MOFLPs.

There are still some clear gaps in the domain of MOFLP research. Scalability to global and real-time contexts, dynamic adaptation to live data, incorporation of multi-stakeholder perspectives, systematic integration of sustainability and resilience, and hybridization with machine learning all represent frontiers for future exploration. Filling in these gaps will not only make facility location models more useful for society, but it will also make them more efficient for computers. The intersection of advanced optimization, digital technologies, and global challenges like climate change, healthcare equity, and sustainable mobility renders facility location research increasingly critical and influential.

In conclusion, the review shows that this is a growing field that is getting bigger, more complex in terms of methods, and more in tune with moral and social needs. Choosing the location of a facility is no longer just a logistical exercise, rather it is now a key decision-making area where efficiency, equity, sustainability, and resilience all come together. Future research, informed by the insights consolidated herein, must persist in merging methodological innovation with practical applicability, thereby guaranteeing that facility location studies continue to serve as a fundamental element of planning and decision-making in intricate, unpredictable, and swiftly evolving contexts.

References

1. Alumur, S.A., Campbell, J.F., Contreras, I., Kara, B.Y., Marianov, V., O'Kelly, M.E.: Perspectives on modeling hub location problems. Eur. J. Oper. Res. **291**(1), 1–17 (2021)
2. Atta, S., Mahapatra, P.R.S., Mukhopadhyay, A.: Deterministic and randomized heuristic algorithms for uncapacitated facility location problem. In: Information and Decision Sciences: Proceedings of the 6th International Conference on FICTA, pp. 205–216. Springer (2018)
3. Atta, S., Mahapatra, P.R.S., Mukhopadhyay, A.: A multi-objective formulation of maximal covering location problem with customers' preferences: exploring pareto optimality-based solutions. Expert Syst. Appl. **186**, 115830 (2021)
4. Ayazi, S., Askarzadeh, A.: A multi-objective cost-loss optimization framework for optimal feeder routing problem of radial distribution networks. Soft. Comput. **28**(6), 5479–5492 (2024)
5. Barbati, M., Corrente, S., Greco, S.: Multiobjective combinatorial optimization with interactive evolutionary algorithms: the case of facility location problems. EURO J. Decis. Process. **12**, 100047 (2024)
6. Bhattacharjee, A.K., Mukhopadhyay, A.: Solving multi-objective hierarchical hub facility location problem using ga-based approaches. In: 2021 IEEE 18th India Council International Conference (INDICON), pp. 1–6. IEEE (2021)
7. Bhattacharjee, A.K., Mukhopadhyay, A.: An improved genetic algorithm with local refinement for solving hierarchical single-allocation hub median facility location problem. Soft Comput. pp. 1–17 (2022)
8. Chen, Y., Lai, Z.: A multi-objective optimization approach for emergency medical service facilities location-allocation in rural areas. Risk Management and Healthcare Policy, pp. 473–490 (2022)
9. Contreras, I., O'Kelly, M.: Hub location problems. Locat. Sci. 327–363 (2019)
10. Demir, İ, Kiraz, B., Ergin, F.C.: Experimental evaluation of meta-heuristics for multi-objective capacitated multiple allocation hub location problem. Eng. Sci. Technol., Int. J. **29**, 101032 (2022)
11. Kahagalage, S., Turan, H.H., Jalalvand, F., El Sawah, S.: A novel graph-theoretical clustering approach to find a reduced set with extreme solutions of pareto optimal solutions for multi-objective optimization problems. J. Global Optim. **86**(2), 467–494 (2023)
12. Khalilzadeh, M., Bahari, A.: A multi-objective mathematical programing model for the problem of p-envy emergency medical service location. Health Serv. Insights **16**, 11786329231195690 (2023)
13. Khodadadi, S., Tasooji, T.K., Shariat-Mohayman, A., Kalantari, N.: A multi-objective simultaneous routing, facility location and allocation model for earthquake emergency logistics. ArXiv preprint arXiv:2503.22487 (2025)
14. Klose, A., Drexl, A.: Facility location models for distribution system design. Eur. J. Oper. Res. **162**(1), 4–29 (2005)
15. Li, K.: Optimizing warehouse logistics scheduling strategy using soft computing and advanced machine learning techniques. Soft. Comput. **27**(23), 18077–18092 (2023)
16. Mokhtarzadeh, M., Tavakkoli-Moghaddam, R., Triki, C., Rahimi, Y.: A hybrid of clustering and meta-heuristic algorithms to solve a p-mobile hub location-allocation problem with the depreciation cost of hub facilities. Eng. Appl. Artif. Intell. **98**, 104121 (2021)
17. Peng, D., Ye, C., Wan, M.: A multi-objective improved novel discrete particle swarm optimization for emergency resource center location problem. Eng. Appl. Artif. Intell. **111**, 104725 (2022) https://doi.org/10.1016/j.engappai.2022.104725, https://www.sciencedirect.com/science/article/pii/S0952197622000379
18. Pourghader Chobar, A., Adibi, M.A., Kazemi, A.: A novel multi-objective model for hub location problem considering dynamic demand and environmental issues. J. Ind. Eng. Manag. Stud. **8**(1), 1–31 (2021)

19. Wang, C., Liu, R., Su, Y., Zhang, X.: Accelerating two-phase multiobjective evolutionary algorithm for electric location-routing problems. IEEE Trans. Intell. Transp. Syst. **24**(9), 9121–9136 (2023)
20. Xue, Y., Li, M., Arabnejad, H., Suleimenova, D., Jahani, A., C Geiger, B., Boesjes, F., Anagnostou, A., JE Taylor, S., Liu, X., et al.: Many-objective simulation optimization for camp location problems in humanitarian logistics (2024)

A Survey on Recent Advances in Multi-objective Evolutionary Clustering

Subham Ghosh, Ujjwal Maulik, and Anirban Mukhopadhyay

Abstract Clustering is a core method in data mining and machine learning that focuses on dividing data into logically consistent and interpretable groups. While traditional clustering methods typically optimize a single-objective function, many real-world problems necessitate the simultaneous consideration of multiple, often conflicting, objectives, such as compactness, separation, and interpretability. Multi-objective clustering addresses this complexity by leveraging evolutionary algorithms and heuristic optimization strategies to generate a diverse set of Pareto-optimal solutions. This chapter presents a comprehensive overview of recent developments in multi-objective clustering, delving into state-of-the-art optimization methods, and challenges in the field. It also examines practical applications across various domains, including image processing, bioinformatics, software engineering, and social network analysis, illustrating the growing relevance and impact of multi-objective approaches in real-world scenarios.

1 Introduction

Clustering [1, 2] is a fundamental technique in data analysis and pattern recognition, playing a pivotal role in uncovering intrinsic structures within datasets without the need for labeled data. It is widely applied across a spectrum of domains, including image and video processing, bioinformatics, financial analytics, social network analysis, and market segmentation. The primary goal of clustering is to partition a dataset into groups, or clusters, such that data points within the same cluster exhibit a high degree of similarity, while those in different clusters are dissimilar from one another. This enables researchers and practitioners to discover hidden structures, patterns, or groupings in data that may not be immediately apparent.

S. Ghosh · U. Maulik
Department of Computer Science and Engineering, Jadavpur University, Kolkata, India

A. Mukhopadhyay (✉)
Department of Computer Science and Engineering, University of Kalyani, Kalyani, India
e-mail: anirban@klyuniv.ac.in

S. Mukhopadhyay et al. (eds.), *Contemporary Advancement in Evolutionary Multi-objective Optimization*, Genetic and Evolutionary Computation,
https://doi.org/10.1007/978-981-95-7908-2_8

Traditional clustering algorithms—such as k-means, hierarchical clustering, and density-based methods like DBSCAN—are generally designed to optimize a single-objective function. This objective is often associated with minimizing intra-cluster distance, maximizing inter-cluster separation, or improving computational efficiency. While these approaches have proven effective in various applications, they often fall short when applied to real-world problems, which are inherently multi-faceted and involve numerous, sometimes conflicting, criteria.

Multi-objective clustering (MOC) has emerged as a promising alternative to address these limitations. By formulating clustering as a multi-objective optimization (MOO) problem, MOC aims to simultaneously optimize multiple conflicting objectives rather than collapsing them into a single aggregated function. The result is not a single clustering solution but a diverse set of Pareto-optimal solutions, each representing a different trade-off among objectives. This approach offers decision-makers greater flexibility to select solutions based on domain-specific requirements and constraints.

To tackle the complexity of MOO in clustering, researchers have employed various optimization strategies, including evolutionary algorithms (such as NSGA-II [3], SPEA2 [4], and MOEA/D [5]), swarm intelligence techniques, and hybrid approaches that integrate local search mechanisms with global exploration. These methods have enabled the development of clustering models that are not only more robust and accurate but also adaptable to high-dimensional, dynamic, or noisy datasets.

Recent research has seen a significant shift toward hybrid and adaptive models, where clustering objectives are dynamically adjusted during the optimization process based on intermediate feedback, data characteristics, or user preferences. Furthermore, advances in deep learning have inspired deep clustering frameworks that incorporate multi-objective optimization principles, allowing deep neural networks to simultaneously learn feature representations and clustering solutions. These techniques have opened new avenues in handling high-dimensional and unstructured data, such as images, texts, and graphs, with remarkable success [6].

Despite the progress made in algorithmic development, multi-objective clustering continues to face several open challenges. One persistent issue in clustering, including MOC, is deciding the optimal number of clusters (K). Another significant issue is the selection of an appropriate solution from the Pareto front. Since the Pareto front can contain a large number of non-dominated solutions, identifying the most suitable one based on domain-specific criteria or user preferences remains a non-trivial task. Another challenge lies in the scalability of MOC algorithms to large-scale and streaming datasets, which demand efficient memory usage and real-time performance. In addition, handling categorical [7], ordinal, or mixed-type data poses substantial difficulties, as most clustering objectives and similarity measures are designed for continuous variables.

Another critical issue is the interpretability of the solutions generated by multi-objective clustering algorithms. While MOC techniques are capable of uncovering nuanced structures in complex data, the resulting models and Pareto fronts are often difficult for end-users to understand and trust—especially in high-stakes applications like healthcare or finance. Addressing this requires the integration of human-in-the-loop approaches, visualization tools, or rule-based interpretability modules that can help translate abstract objective trade-offs into actionable insights.

This chapter provides a comprehensive overview of recent advances in multi-objective clustering. We explore the foundational principles of MOC, survey state-of-the-art algorithms and optimization frameworks and we will see how these optimization algorithms can be effectively used in optimizing multiple objectives of a clustering process. Finally, we present a set of real-world applications that demonstrate the practical value of multi-objective clustering in fields such as bioinformatics, social network analysis, cybersecurity, and environmental modeling.

2 Clustering

Clustering is one of the most important tasks in unsupervised learning. It helps us to automatically group similar data points together without using any predefined class labels. Thus, clustering helps to understand the underlying structure of data and discover hidden patterns.

2.1 *Formal Definition*

Let $X = \{x_1, x_2, \dots, x_N\}$ denote a dataset of N data points, where each $x_i \in R^d$ represents a d-dimensional feature vector. The goal of clustering is to partition X into K non-overlapping subsets (clusters) $C = \{C_1, C_2, \dots, C_K\}$ such that:

$$\bigcup_{k=1}^{K} C_k = X, \quad C_i \cap C_j = \emptyset, \quad \forall i \neq j. \tag{1}$$

Each cluster C_k is represented by a prototype or centroid $\boldsymbol{\mu}_k$, often computed as the mean of the points assigned to it:

$$\boldsymbol{\mu}_k = \frac{1}{|C_k|} \sum_{x_i \in C_k} x_i. \tag{2}$$

2.2 *What Makes a Good Clustering?*

A good clustering should have:

- **Compactness**: Points in the same cluster should be close together.
- **Separation**: Different clusters should be far apart.

Many algorithms try to find a balance between these two goals. Traditionally, most clustering algorithms optimize a single objective such as minimizing the overall distance between data points and their cluster centers:

$$J = \sum_{i=1}^{k} \sum_{x_j \in C_i} \|x_j - \mu_i\|^2,$$

where μ_i is the center (mean) of cluster C_i. However, as we will see later, this single-objective approach may not always give the best clustering for complex data, leading to the idea of *multi-objective clustering*.

2.3 Major Types of Clustering Methods

Different clustering algorithms approach the problem in different ways. Each type has its own assumptions and strengths. Let us discuss the most popular ones.

Partition-Based Clustering

The most common example is the k-**means** algorithm [8]. It works by iteratively assigning points to the nearest cluster center and then recalculating the centers. The aim is to minimize the total squared distance of points from their cluster center:

$$J = \sum_{i=1}^{k} \sum_{x_j \in C_i} \|x_j - \mu_i\|^2.$$

The algorithm is simple:

1. Choose k initial cluster centers (randomly or using k-means++ initialization).
2. Assign each data point to its nearest center.
3. Recalculate the centers as the mean of points in each cluster.
4. Repeat steps 2–3 until the centers stop changing.

Although easy to implement and very fast, k-means has some drawbacks: it assumes clusters are roughly spherical and of similar size, it is sensitive to noise, and the user must choose k beforehand.

A more robust variant is k-**medoids** (or PAM–Partitioning Around Medoids) [9]. Instead of using the mean, it selects an actual data point (called a medoid) as the cluster center. The cost function is slightly different:

$$J = \sum_{i=1}^{k} \sum_{x_j \in C_i} \|x_j - m_i\|,$$

where m_i is the medoid of cluster C_i. This makes k-medoids less sensitive to outliers.

Hierarchical Clustering

Hierarchical clustering builds a tree-like structure (called a dendrogram) showing how clusters are formed step by step [10]. There are two strategies:

- **Agglomerative (bottom-up):** Start with each data point as its own cluster and merge the two closest clusters at each step.
- **Divisive (top-down):** Start with all points in one cluster and split it gradually into smaller ones.

The key difference lies in how "closeness" between clusters is measured:

$$\text{Single linkage: } D(C_i, C_j) = \min_{x \in C_i, y \in C_j} \|x - y\|,$$

$$\text{Complete linkage: } D(C_i, C_j) = \max_{x \in C_i, y \in C_j} \|x - y\|,$$

$$\text{Average linkage: } D(C_i, C_j) = \frac{1}{|C_i||C_j|} \sum_{x \in C_i} \sum_{y \in C_j} \|x - y\|.$$

The result is a visual tree that can be "cut" at different levels to produce a desired number of clusters. This approach is intuitive but computationally heavy for large datasets.

Density-Based Clustering

Density-based methods, such as **DBSCAN** [11], are based on a simple idea: clusters are dense areas of data points separated by regions of low density.

DBSCAN uses two parameters:

- ε—the radius of the neighborhood around a point.
- MinPts—the minimum number of points required to form a dense region.

A point is considered a *core point* if at least MinPts other points are within its ε-neighborhood. Clusters grow by connecting core points that are close to each other. Unlike k-means, DBSCAN can find clusters of arbitrary shape and detect noise, but it can struggle if the data have varying densities.

Model-Based Clustering

Model-based methods assume that data come from a mixture of underlying probability distributions. A popular example is the **Gaussian Mixture Model (GMM)** [12].

It assumes that each cluster can be represented as a Gaussian (normal) distribution. GMMs are flexible and can model overlapping clusters, but they require good initialization and are computationally more demanding.

Grid-Based and Graph-Based Clustering

Grid-based methods divide the data space into cells and group dense cells together. These methods are very fast for large datasets. Graph-based methods, such as spectral clustering, view the data as a graph where each node is a data point and edges represent similarities. The algorithm finds the best way to cut the graph into parts so that the connections between different clusters are minimal while internal connections remain strong.

2.4 Challenges in Clustering

Despite its wide applicability, clustering faces several challenges:

1. **Selection of the number of clusters** (k): Many algorithms require the number of clusters as input, which is often unknown in real-world data.
2. **High-dimensional and heterogeneous data:** As the dimensionality increases, traditional distance measures (e.g., Euclidean) become less meaningful, leading to the "curse of dimensionality."
3. **Scalability and noise sensitivity:** Large-scale datasets and noisy observations can adversely affect the clustering performance.
4. **Multiple conflicting criteria:** A single-objective approach may fail to balance compactness, separation, connectivity, and interpretability simultaneously.

These limitations motivate the development of *multi-objective clustering* techniques, which aim to optimize multiple conflicting objectives concurrently rather than a single one. By exploring the trade-offs among different clustering criteria, multi-objective algorithms can yield a set of Pareto-optimal solutions, offering greater flexibility and robustness in clustering analysis. The next section introduces this idea in detail and explains how multi-objective optimization brings new power and flexibility to clustering.

3 Multi-objective Optimization

Multi-objective Optimization (MOO) is used to maximize/minimize two or more objectives simultaneously.

3.1 Introduction to Multi-objective Optimization

A general multi-objective optimization problem (MOP) can be formulated as:

$$\text{Minimize/Maximize} \quad \mathbf{F}(\mathbf{x}) = [f_1(\mathbf{x}), f_2(\mathbf{x}), \ldots, f_m(\mathbf{x})], \tag{3}$$

subject to:

$$\mathbf{x} \in \Omega, \tag{4}$$

where $\mathbf{x}$ is a solution vector in the decision space Ω, and $\mathbf{F}(\mathbf{x})$ represents m objective functions.

A solution $\mathbf{x}_1$ is said to *dominate* another solution $\mathbf{x}_2$ (denoted as $\mathbf{x}_1 \prec \mathbf{x}_2$) if:

$$\forall i, \; f_i(\mathbf{x}_1) \leq f_i(\mathbf{x}_2) \quad \text{and} \quad \exists j : f_j(\mathbf{x}_1) < f_j(\mathbf{x}_2). \tag{5}$$

The set of all non-dominated solutions forms the *Pareto-optimal set*, and the corresponding points in the objective space form the *Pareto front*. Unlike single-objective optimization, where one optimal solution exists, MOO aims to approximate this Pareto front.

Evolutionary algorithms are particularly well-suited for MOO because they work with a population of solutions and can explore multiple trade-offs simultaneously. The three most prominent evolutionary multi-objective algorithms are NSGA-II, SPEA2, and MOEA/D, which are explained in detail below.

3.2 Non-dominated Sorting Genetic Algorithm II (NSGA-II)

The Non-dominated Sorting Genetic Algorithm II (NSGA-II), developed by Deb et al. (2002) [3], is a widely used evolutionary algorithm that improves upon the original NSGA by introducing:

- Fast non-dominated sorting (to classify individuals by Pareto dominance),
- Elitism (preserving the best solutions across generations),
- Crowding distance (to maintain diversity among solutions).

Step-by-Step Procedure

1. **Initialization**: Generate an initial random population P_0 of size N.
2. **Evaluation**: Compute the values of all objective functions $f_1, f_2, \ldots, f_m$ for each individual.
3. **Non-dominated Sorting**:
   - Classify the population into Pareto fronts $F_1, F_2, \ldots$.
   - Individuals in F_1 are not dominated by any other individual; those in F_2 are dominated only by members of F_1, and so on.

4. **Crowding Distance Calculation:** For individuals in the same front, calculate a crowding distance that estimates how close a solution is to its neighbors:

$$d_i = \sum_{k=1}^{m} \frac{f_k(i+1) - f_k(i-1)}{f_k^{\max} - f_k^{\min}}. \tag{6}$$

   Larger d_i means more isolated (diverse) solutions.
5. **Selection**: Perform binary tournament selection based on *rank* (front number) and *crowding distance*.
6. **Crossover and Mutation**: Apply standard genetic operators to generate offspring population Q_t.
7. **Combine and Sort**: Combine parent and offspring populations to form $R_t = P_t \cup Q_t$ and perform non-dominated sorting.
8. **Next Generation**: Fill the next population P_{t+1} by choosing individuals starting from the best front until population size N is reached.

Key Advantages

- Maintains a good balance between convergence and diversity.
- Efficient sorting with $O(MN^2)$ time complexity.
- Suitable for clustering where multiple objectives (compactness, separation, balance) are optimized simultaneously.

3.3 *Strength Pareto Evolutionary Algorithm 2 (SPEA2)*

SPEA2, proposed by Zitzler et al. (2001) [4], improves upon the original SPEA algorithm with a refined fitness assignment mechanism, an external elitist archive, and a density estimation technique for diversity preservation.

Step-by-Step Procedure

1. **Initialization:** Create an initial random population P_0 and an empty archive A_0.
2. **Fitness Assignment:** Each individual i is assigned a *strength value* $S(i)$ equal to the number of solutions it dominates:

$$S(i) = |\{j | i \prec j\}|. \tag{7}$$

   The raw fitness $R(i)$ is computed as the sum of strengths of solutions dominating i:

$$R(i) = \sum_{j \prec i} S(j). \tag{8}$$

   A density term $D(i)$ is then added to balance crowding:

$$D(i) = \frac{1}{\sigma_i^k + 2}, \tag{9}$$

where σ_i^k is the distance to the k-th nearest neighbor. The final fitness is:

$$F(i) = R(i) + D(i). \tag{10}$$

3. **Archive Update:** Combine P_t and A_t, select the best non-dominated individuals based on fitness to form A_{t+1}.
4. **Selection:** Use individuals from the archive to generate the mating pool.
5. **Crossover and Mutation:** Generate offspring population using standard genetic operators.
6. **Replacement:** The new population P_{t+1} is formed from offspring, and the archive is updated accordingly.

Key Advantages

- Ensures elitism by using an external archive.
- Provides fine-grained fitness assignment and density control.
- Produces a well-distributed and stable Pareto front.

3.4 Multi-objective Evolutionary Algorithm Based on Decomposition (MOEA/D)

The Multi-objective Evolutionary Algorithm based on Decomposition (MOEA/D), proposed by Zhang and Li (2007) [5], approaches multi-objective optimization differently. Instead of comparing solutions using dominance, it decomposes the MOP into several scalar subproblems, each corresponding to a weight vector, and solves them simultaneously.

Mathematical Concept Each subproblem minimizes a scalar aggregation function, often the Tchebycheff function:

$$g^{te}(\mathbf{x}|\boldsymbol{\lambda}, \mathbf{z}^*) = \max_{1 \le i \le m} \lambda_i |f_i(\mathbf{x}) - z_i^*|, \tag{11}$$

where

- $\boldsymbol{\lambda} = (\lambda_1, \lambda_2, \ldots, \lambda_m)$ is a weight vector,
- $\mathbf{z}^* = (z_1^*, z_2^*, \ldots, z_m^*)$ is the ideal point (best known value for each objective).

Step-by-Step Procedure

1. **Initialization:** Define N weight vectors $\boldsymbol{\lambda}^1, \boldsymbol{\lambda}^2, \ldots, \boldsymbol{\lambda}^N$ uniformly on the simplex.
2. **Neighborhood Formation:** For each subproblem i, define a neighborhood consisting of the T closest weight vectors.
3. **Population Initialization:** Randomly generate N solutions, one for each subproblem.

4. **Evaluation:** Calculate $\mathbf{F}(\mathbf{x})$ and update the reference point $\mathbf{z}^*$.
5. **Reproduction:** For each subproblem, select parents from its neighborhood and apply crossover and mutation to generate offspring $\mathbf{y}$.
6. **Update Neighboring Solutions:** If $\mathbf{y}$ improves any neighboring subproblem's objective (based on the scalar function g^{te}), replace that neighbor's solution with $\mathbf{y}$.
7. **Termination:** Repeat until the maximum number of generations is reached.

Key Advantages

- Highly scalable for problems with many objectives.
- Efficient neighborhood-based search encourages cooperation among subproblems.
- Maintains a well-distributed Pareto front naturally through decomposition.

In summary, multi-objective optimization provides a comprehensive framework for addressing conflicting clustering objectives. NSGA-II emphasizes rank-based selection and diversity, SPEA2 refines elitism and density estimation, while MOEA/D introduces a decomposition-based perspective. Together, these algorithms form the foundation of modern multi-objective clustering methods that can effectively balance compactness, separation, and structural stability in high-dimensional data.

4 Multi-objective Clustering

Multi-objective clustering (MOC) is a modern clustering paradigm that addresses the limitations of traditional single-objective clustering algorithms by simultaneously optimizing multiple, and often conflicting, objectives. This approach enables the discovery of diverse data structures that may not be captured by optimizing a single criterion alone. For instance, an MOC algorithm that jointly considers compactness-based and connectedness-based criteria can effectively detect various types of data distributions, as illustrated in Fig. 1. In contrast, algorithms relying solely on a compactness-based criterion, such as k-means (KM), are capable of identifying only globular clusters (Fig. 1a) but fail to detect non-convex or ring-shaped clusters (Fig. 1b) and heterogeneous structures (Fig. 1c). Conversely, connectedness-based algorithms, such as Shared Nearest Neighbor (SNN), can successfully identify ring-shaped structures (Fig. 1b) but struggle with globular and heterogeneous patterns (Fig. 1a, c). Therefore, MOC frameworks provide a unified mechanism to balance multiple clustering quality measures, resulting in a more comprehensive understanding of complex and heterogeneous datasets.

To achieve this power of detecting complex clustering structure, evolutionary multi-objective algorithms, such as NSGA-II, SPEA2, and MOEA/D, are employed to evolve a population of clustering solutions toward a Pareto-optimal set that represents the best trade-offs among the objectives. AMOSA [13], NSGA-III [14, 15] are some more recent advanced multi or many objective optimization algorithms that could also be employed in the domain of MOC.

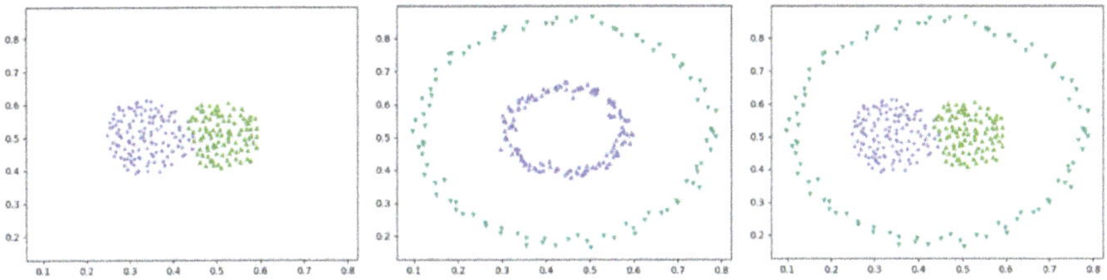

Fig. 1 Different data structures (objects with the same color represent a cluster in each sub-figure): **a** globular clusters, **b** ring-shaped clusters, and **c** heterogeneous data structure.

4.1 Mathematical Formulation

Let $X = \{\mathbf{x}_1, \mathbf{x}_2, \ldots, \mathbf{x}_n\}$ be a dataset of n data points in d-dimensional space, i.e., $\mathbf{x}_i \in \mathbb{R}^d$. The goal of clustering is to partition X into K clusters, $C = \{C_1, C_2, \ldots, C_K\}$, where

$$\bigcup_{k=1}^{K} C_k = X, \quad C_i \cap C_j = \emptyset \text{ for } i \neq j. \tag{12}$$

In a multi-objective setting, we optimize multiple clustering quality indices simultaneously, such as:

1. **Compactness:** Minimizing intra-cluster variance

$$f_1 = \sum_{k=1}^{K} \frac{1}{|C_k|} \sum_{\mathbf{x}_i \in C_k} \|\mathbf{x}_i - \boldsymbol{\mu}_k\|^2, \tag{13}$$

   where $\boldsymbol{\mu}_k$ is the centroid of cluster C_k.
2. **Separation:** Maximizing inter-cluster distances

$$f_2 = -\frac{2}{K(K-1)} \sum_{i=1}^{K} \sum_{j=i+1}^{K} \|\boldsymbol{\mu}_i - \boldsymbol{\mu}_j\|. \tag{14}$$

3. **Balance:** Minimizing cluster size variance

$$f_3 = \text{Var}(|C_1|, |C_2|, \ldots, |C_K|). \tag{15}$$

Thus, the problem becomes:

$$\text{Minimize } \mathbf{F}(\mathbf{S}) = [f_1(\mathbf{S}), f_2(\mathbf{S}), f_3(\mathbf{S})], \tag{16}$$

where **S** denotes a clustering solution (i.e., a specific partition of the dataset). Any cluster validity index [16–23] could also be used as an objective to be optimized by the multi-objective optimization algorithms.

4.2 Solution Representation

Each individual (solution) in the population represents a possible partitioning of the dataset. Common representations include:

- **Centroid-Based Encoding:** Each chromosome encodes K centroids $\mu_1, \mu_2, \ldots, \mu_K$. Each data point is assigned to the nearest centroid.
- **Label-Based Encoding:** Each chromosome has n genes, where the i-th gene indicates the cluster label of data point $\mathbf{x}_i$.

Label-based representation is more direct but can be redundant for large datasets. Centroid-based encoding is often preferred for continuous feature spaces.

4.3 Algorithmic Framework

A general framework for multi-objective clustering based on evolutionary algorithms is presented below.

Algorithm 1 General Framework for Multi-objective Clustering

```
Input: Dataset X, number of clusters K, population size N, number of generations G.
Output: Pareto-optimal set of clustering solutions.
Initialize population P_0 with N random clustering solutions.
for t = 1 to G do
    Evaluate objective functions F(S) = [f_1, f_2, ..., f_m] for each S ∈ P_t.
    Perform non-dominated sorting to classify individuals into Pareto fronts.
    Calculate diversity measure (e.g., crowding distance or density).
    Select parents for reproduction based on rank and diversity.
    Apply crossover and mutation to generate offspring population Q_t.
    Combine P_t and Q_t into R_t = P_t ∪ Q_t.
    Select the best N individuals from R_t using Pareto dominance and diversity preservation.
    Update the reference archive of non-dominated solutions.
end for
Return: The non-dominated set representing the Pareto front.
```

4.4 Initialization

The algorithm starts with a randomly generated population of N clustering solutions. For centroid-based encoding, centroids can be initialized by randomly selecting K data points from X. For label-based encoding, each data point is assigned a random cluster label from $\{1, 2, \ldots, K\}$.

Optionally, one may use the output of traditional clustering algorithms such as k-means or fuzzy c-means to seed part of the initial population to accelerate convergence.

4.5 Evaluation of Objective Functions

For each clustering solution $\mathbf{S}$:

1. Compute the centroids $\boldsymbol{\mu}_k$ for all clusters.
2. Evaluate the objective functions (e.g., compactness, separation, and balance).
3. Normalize objective values if necessary to ensure consistent scaling.

Each solution is then ranked based on Pareto dominance, forming multiple fronts $F_1, F_2, \ldots$. Solutions in F_1 are non-dominated and represent the best trade-offs found so far.

4.6 Genetic Operators

Genetic operations are the core components that drive exploration and exploitation in the search space.

Selection Selection is usually performed using binary tournament selection, where two individuals are compared based on their rank and diversity measure (e.g., crowding distance). The one with a better trade-off is selected as a parent.

Crossover Crossover combines information from two parent solutions to generate offspring. Common crossover strategies include:

- **Arithmetic Crossover (for centroid encoding):**

$$\boldsymbol{\mu}_k^{(c)} = \alpha \boldsymbol{\mu}_k^{(p_1)} + (1 - \alpha) \boldsymbol{\mu}_k^{(p_2)}, \quad \alpha \in [0, 1]. \tag{17}$$

- **Uniform Crossover (for label encoding):** Each gene (cluster label) is selected from either parent with equal probability.

Mutation Mutation introduces randomness to maintain diversity. For centroid encoding:

$$\boldsymbol{\mu}_k' = \boldsymbol{\mu}_k + \epsilon, \quad \epsilon \sim \mathcal{N}(0, \sigma^2), \tag{18}$$

where σ is chosen empirically or adaptively depending on the problem.

For label encoding, a small proportion of data points may be reassigned to random clusters.
Elitism and Replacement After generating offspring, parent and offspring populations are merged, and the best N individuals are retained based on non-dominated sorting and diversity measures. This ensures that the best solutions are never lost across generations.

4.7 Termination and Pareto Front Extraction

The algorithm terminates when the maximum number of generations (G) is reached or when convergence criteria (e.g., negligible improvement in Pareto front) are satisfied. The final output is the Pareto-optimal set of clustering solutions:

$$P^* = \{\mathbf{S}_1, \mathbf{S}_2, \ldots, \mathbf{S}_p\}, \tag{19}$$

where each $\mathbf{S}_i$ represents a different trade-off among the objectives. The decision-maker can then choose a suitable clustering solution based on domain-specific requirements, such as prioritizing compactness or balance.

4.8 Advantages of Multi-objective Clustering

- Provides a set of trade-off clustering solutions rather than a single output.
- Avoids the need for weight-based combination of objectives.
- Can adapt to various clustering criteria (compactness, separation, silhouette, balance).
- Leverages the exploration capability of evolutionary algorithms to escape local minima.

In conclusion, multi-objective clustering bridges the gap between unsupervised learning and evolutionary multi-objective optimization. By optimizing multiple conflicting cluster quality metrics simultaneously, it produces diverse, high-quality clustering solutions that reflect the inherent trade-offs present in real-world data analysis tasks.

5 Challenges in Multi-Objective Clustering

Despite its ability to uncover diverse and complex cluster structures, multi-objective clustering (MOC) still faces several challenges that limit its practical applicability and performance. These challenges arise from both the optimization process and the nature of real-world data. Recent studies have made significant progress in addressing these issues through innovative algorithmic designs, hybrid models, and adaptive mechanisms. This section summarizes the major challenges in MOC and outlines how recent research efforts have attempted to overcome them.

5.1 Determining the Number of Clusters

One persistent issue in clustering, including MOC, is deciding the optimal number of clusters (K). Setting K too low leads to over-generalization, while a large K results in fragmented clusters. Some recent MOC methods treat K as a variable rather than a fixed parameter, allowing the evolutionary process to evolve both the cluster assignments and the number of clusters simultaneously. For example, adaptive chromosome encoding techniques and automatic cluster merging/splitting operators have been proposed to dynamically adjust K during optimization.

Evolutionary multi-objective-based methods exploit their inherent parallelism to simultaneously explore multiple clustering solutions with different k values. In the study titled *"Multi-clustering via evolutionary multi-objective optimization,"* [24] a bi-objective model was proposed where both the number of clusters and the sum of squared distances (SSD) between data points and their centroids were treated as optimization objectives. By applying a novel transformation strategy to make these objectives conflicting, and using the NSGA-II algorithm to search for Pareto-optimal solutions, the approach—termed EMO-k-clustering—was able to generate a diverse set of clustering results in a single run. A validity index was then used to select the most suitable clustering from the obtained Pareto front. Experimental results on multiple datasets demonstrated that EMO-k-clustering is both effective and efficient, providing a promising direction for automatic and adaptive clustering without prior knowledge of k.

5.2 High Computational Complexity

Evolutionary algorithms used in MOC are computationally intensive because they evaluate multiple objectives for each individual solution across several generations. As the dataset size or number of objectives increases, the computational cost can grow rapidly. Recent works have attempted to mitigate this problem through various means: using parallel and distributed implementations, surrogate models to approximate expensive evaluations, and adaptive population management strategies that reduce redundant computations while preserving diversity.

The evolutionary multi-objective k-clustering (EMO-KC) framework explained in the previous subsection titled "Determining the Number of Clusters" exhibits limited clustering accuracy on high-dimensional data and may produce invalid solutions during initialization due to ineffective optimization of the transformed objective (SSD_{exp}). To overcome these limitations, the study titled *"An Improved Evolutionary Multi-Objective Clustering Algorithm Based on Autoencoder"* proposed AE-IEMOKC [6], which integrates an autoencoder with an improved version of EMO-KC (IEMO-KC). The enhanced approach introduces a scaling factor to better optimize SSD_{exp} and a valid initialization mechanism to prevent invalid solutions. Experimental results demonstrated that AE-IEMOKC achieves higher accuracy and better validity, highlighting the importance of hybrid approaches that combine deep learning and evolutionary optimization to address the challenges of accuracy, convergence, and robustness in multi-objective clustering.

5.3 Data-Driven Adaptation and Objective Function Challenges in Multi-Objective Clustering

A recent advancement in the field of multi-objective clustering is the development of adaptive frameworks that consider the intrinsic properties of data to guide the optimization process. In the study titled *"An Adaptive Evolutionary Multi-Objective Clustering Based on the Data Properties of the Base Partitions,"* [25] the authors addressed persistent challenges in evolutionary multi-objective clustering (EMOC), such as the proper definition of objective functions and effective initialization strategies. These challenges often lead to inadequate optimization or restricted search behavior. To overcome this, the authors proposed the **data separation degree (DSD)**, a novel unsupervised metric designed to evaluate the separation or overlap among clusters in a dataset. By integrating DSD with an adaptive multi-objective framework, termed **AEMOC** (Adaptive Evolutionary Multi-Objective Clustering), the approach enables the selection of objective functions and algorithmic parameters based on data characteristics. Unlike traditional EMOC methods that use a fixed set of objectives, AEMOC dynamically adapts its optimization strategy according to the data's degree of separation or overlap. Experimental evaluations on artificial and real-world datasets demonstrated that AEMOC not only improves clustering accuracy but also enhances robustness and generalization by aligning optimization goals with data properties, representing a significant step forward in adaptive and intelligent multi-objective clustering.

5.4 Dealing with Conflicting Information from Multiple Data Views

In many real-world problems, the same data can be described from several different points of view. For example, in bioinformatics, a protein can be described by its sequence, its structure, or its chemical properties. In social networks, a person might be represented by their connections, messages, and interests. Each of these descriptions is called a *data view*. When we try to cluster such data, it becomes difficult because each view may show different or even conflicting patterns. This situation is known as *multi-view clustering*.

Traditional multi-objective clustering (MOC) algorithms usually work with only one view of the data. They assume that all the useful information is contained in a single set of features. However, when data come from multiple views, this assumption often fails. Some patterns might appear clearly in one view but not in another, and combining these conflicting sources of information becomes a major challenge.

To address this issue, the paper titled *"Evolutionary Multi-Objective Clustering Over Multiple Conflicting Data Views"* [26] proposed a new evolutionary approach that can handle data from different views at the same time. The method represents the data using a graph-based structure built with a minimum spanning tree (MST) and neighborhood relationships. This structure helps preserve local data patterns without forcing the algorithm to form specific cluster shapes. The algorithm then treats each data view as a separate objective and optimizes them all together. In this way, it searches for clustering solutions that balance the conflicting information across the views.

Experiments on both synthetic and real-world datasets, including a bioinformatics task of grouping similar protein structures, showed that the proposed method can automatically find the right number of clusters and handle noisy or unreliable data sources. The results demonstrated that considering multiple data views at once can lead to more accurate and robust clustering outcomes. This research highlights an important challenge in MOC—how to combine and balance multiple, possibly conflicting, sources of information to obtain meaningful clusters.

5.5 *Simultaneous Feature Selection and Clustering*

In many real-world datasets, not all features (or variables) are equally useful for clustering. Some features may contain noise, redundant information, or irrelevant details that can confuse the clustering algorithm and reduce its performance. For example, in a dataset describing customer behavior, features like "customer ID" or "transaction time" may not contribute meaningfully to identifying customer groups. Therefore, selecting a proper subset of informative features is an important step in improving clustering quality.

However, finding the best set of features and the best cluster structure at the same time is a challenging problem. Traditional approaches usually perform feature selection first and then apply clustering on the selected features. But this two-step approach may not always work well because the best features for clustering often depend on the cluster structure itself. To solve this dependency, researchers have developed techniques for **simultaneous feature selection and clustering** (SFSC), where both processes are optimized together.

In a multi-objective clustering (MOC) framework, this problem can be naturally modeled as an optimization task with multiple conflicting goals. The main objectives are usually:

- Minimizing clustering error or maximizing cluster compactness and separation.
- Minimizing the number of selected features while keeping enough information to form meaningful clusters.

Mathematically, this can be expressed as a multi-objective optimization problem:

$$\text{Minimize } F(x, S) = \big(f_1(x, S), f_2(x, S)\big),$$

where

- x represents the cluster assignments of data points,
- $S \subseteq \{1, 2, \ldots, d\}$ represents the subset of selected features from the full feature set of dimension d,
- $f_1(x, S)$ measures the clustering quality (e.g., within-cluster variance), and
- $f_2(x, S)$ measures the feature subset size or redundancy.

Evolutionary algorithms are well-suited for this type of problem because they can explore large and complex search spaces effectively. Each individual in the population can represent both a cluster partition and a feature subset. For example, a binary string can be used to indicate which features are selected, while another part of the representation encodes the cluster assignments. The algorithm then searches for Pareto-optimal solutions that balance between using fewer features and achieving better clustering results.

Simultaneous feature selection and clustering [27, 28] provides several benefits:

- It reduces the effect of noisy and irrelevant features.
- It improves the interpretability of clusters by using fewer, more meaningful features.
- It enhances computational efficiency by working in a smaller feature space.

However, SFSC also increases the complexity of the optimization problem because both clustering and feature selection must be solved together. As a result, designing efficient algorithms that can balance these conflicting objectives and scale to high-dimensional data remains an active research challenge in multi-objective clustering.

5.6 Interpretability and Validation of Results

Since MOC produces multiple Pareto-optimal clustering solutions, selecting the most meaningful one for practical use can be difficult. Interpreting and validating these results remains an open challenge. Some recent research has introduced automatic cluster validity indices or visualization-based selection tools to help decision-makers choose the most suitable clustering. Others propose integrating domain knowledge into the optimization process, allowing users to guide the search toward interpretable and application-relevant solutions.

6 Applications of Multi-objective Clustering

Multi-objective clustering has gained significant attention in recent years due to its ability to optimize multiple conflicting criteria simultaneously, making it particularly suitable for complex real-world problems. In recent times, multi-objective clustering has been applied in various domains like image segmentation, bioinformatics, web mining, social networks, software engineering, and time-series analysis. Some of the important applications are discussed below.

In image segmentation, Zhao et al. [29] used an ensemble CART surrogate-assisted multi-objective evolutionary model to generate Pareto-optimal segmentation results by balancing intra-region compactness and boundary preservation. In bioinformatics, multi-objective clustering has gained significant attention for its ability to extract biologically meaningful patterns from high-dimensional data [30]. Rodríguez-Bejarano et al. [31] introduced RVEA-CAST, a reference vector-guided evolutionary algorithm for clustering single-cell transcriptomic data, where conflicting objectives such as cluster purity, connectivity, and population balance are optimized simultaneously to reveal cell heterogeneity. In web mining, Sánchez-Gómez et al. [32] applied multi-objective swarm intelligence to improve document clustering by jointly maximizing inter-cluster separation and intra-cluster cohesion, while Su et al. [33] addressed scalable social network analysis using a parallel multi-objective evolutionary algorithm for community detection in large-scale complex graphs. In software engineering, Yuste et al. [34] optimized software modularity using a multi-objective variable neighborhood search approach that restructures module dependency graphs by minimizing coupling and maximizing cohesion—providing multiple architectural refactoring options along the Pareto front. For time-series analysis, Korlakov et al. [35] treated dynamic clustering as a multi-objective task, integrating temporal consistency with clustering accuracy to select evolving partitions across time. Beyond these domains, multi-objective clustering has also been explored in computer security, where it enables the design of intrusion detection systems that balance competing goals such as detection rate, false-alarm minimization, and computational efficiency, demonstrating its versatility in modern intelligent systems.

7 Conclusion

Multi-objective clustering (MOC) is an important approach that helps us solve clustering problems where more than one goal needs to be considered at the same time.

In traditional clustering methods, we usually focus on one objective, such as keeping data points in each cluster as close as possible. However, in real-world data, different objectives may conflict with each other. For example, making clusters compact may reduce how well they are separated from one another. MOC helps us handle such conflicting goals by finding a set of balanced solutions rather than just one.

The use of ideas from multi-objective optimization (MOO) has made MOC more powerful. Algorithms such as NSGA-II, SPEA2, and MOEA/D are commonly used in this field. These algorithms do not look for a single best clustering result. Instead, they generate a group of good solutions, known as the Pareto front, which shows different trade-offs among the objectives. This allows researchers and practitioners to choose the most suitable solution based on their needs.

Although MOC has shown great success, there are still many challenges. It is often difficult to define the right set of objective functions that truly describe what a good clustering should look like. Another problem is balancing two important goals: finding high-quality solutions (convergence) and keeping a wide variety of solutions (diversity). In addition, as datasets become larger and more complex, we need algorithms that can scale efficiently. Evaluating the quality of multi-objective clustering solutions using proper cluster validity indices is also an ongoing research area.

MOC has been applied in many fields, such as bioinformatics, image processing, social network analysis, and market research. In these areas, MOC helps reveal different useful patterns and provides a more complete understanding of the data. This flexibility makes MOC very suitable for solving modern, real-world data problems.

Looking toward the future, research in MOC is likely to move toward combining it with deep learning, hybrid optimization techniques, and visualization tools that make it easier to understand the trade-offs among solutions. Another important direction is to make MOC more interpretable and to include expert knowledge during the clustering process. These improvements will make MOC more useful for real-world decision-making.

In summary, multi-objective clustering is a growing and exciting research area that connects optimization and data analysis. Its continued development will help us extract meaningful and balanced insights from the huge and complex data we face today.

References

1. Jain, A.K., Murty, M.N., Flynn, P.J.: Data clustering: a review. ACM Comput. Surv. (CSUR) **31**(3), 264–323 (1999)
2. Dubes, R., Jain, A.K.: Validity studies in clustering methodologies. Pattern Recogn. **11**(4), 235–254 (1979)
3. Deb, K., Pratap, A., Agarwal, S., Meyarivan, T.: A fast and elitist multiobjective genetic algorithm: NSGA-II. IEEE Trans. Evol. Comput. **6**(2), 182–197 (2002)
4. Zitzler, E., Laumanns, M., Thiele, L.: SPEA2: improving the strength pareto evolutionary algorithm. TIK-Report 103. ETH Zurich, Switzerland (2001)

5. Zhang, Q., Li, H.: Moea/d: a multiobjective evolutionary algorithm based on decomposition. IEEE Trans. Evol. Comput. **11**(6), 712–731 (2007)
6. Qiu, M., Zhang, Y., Lei, S., Gu, M.: An improved evolutionary multi-objective clustering algorithm based on autoencoder. Appl. Sci. **14**(6), 2454 (2024)
7. Mukhopadhyay, A., Maulik, U.: Multiobjective approach to categorical data clustering. In: IEEE Congress on Evolutionary Computation, pp. 1296–1303. IEEE (2007)
8. MacQueen, J.: Some methods for classification and analysis of multivariate observations. In: Le Cam, L.M., Neyman, J. (eds.), Proceedings of the Fifth Berkeley Symposium on Mathematical Statistics and Probability, vol. 1, pp. 281–297. University of California Press (1967)
9. Kaufman, L., Rousseeuw, P.J.: Clustering by means of medoids. In: Dodge, Y. (ed.) Statistical data analysis based on the L1 norm and related methods, pp. 405–416. North-Holland (1987)
10. Johnson, S.C.: Hierarchical clustering schemes. Psychometrika **32**(3), 241–254 (1967)
11. Ester, M. Kriegel, H.-P., Sander, J., Xu, X.: A density-based algorithm for discovering clusters in large spatial databases with noise. In: Proceeding of 2nd International Conference Knowledge Discovery and Data Mining (KDD-96), pp. 226–231 (1996)
12. McLachlan, G.J., Peel, D.: Finite Mixture Models. Wiley, New York (2000)
13. Bandyopadhyay, S., Saha, S., Maulik, U., Deb, K.: A simulated annealing-based multiobjective optimization algorithm: AMOSA. IEEE Trans. Evol. Comput. **12**(3), 269–283 (2008)
14. Deb, K., Jain, H.: An evolutionary many-objective optimization algorithm using reference-point-based nondominated sorting approach, Part I: Solving problems with box constraints. IEEE Trans. Evol. Comput. **18**(4), 577–601 (2014)
15. Jain, H., Deb, K.: An evolutionary many-objective optimization algorithm using reference-point-based nondominated sorting approach, Part II: Handling constraints and extending to an adaptive approach. IEEE Trans. Evol. Comput. **18**(4), 602–622 (2014)
16. Davies, D.L., Bouldin, D.W.: A cluster separation measure. IEEE Transactions on Pattern Analysis and Machine Intelligence, vol. PAMI-1, no. 2, pp. 224–227 (1979)
17. Dunn, J.C.: Well-separated clusters and optimal fuzzy partitions. J. Cybernet. **4**(1), 95–104 (1974)
18. Rousseeuw, P.J.: Silhouettes: a graphical aid to the interpretation and validation of cluster analysis. J. Comput. Appl. Math. **20**, 53–65 (1987)
19. Xie, X.L., Beni, G.: A validity measure for fuzzy clustering. IEEE Trans. Pattern Anal. Mach. Intell. **13**(8), 841–847 (1991)
20. Pakhira, P.K., Bandyopadhyay, S., Maulik, U.: Validity index for crisp and fuzzy clusters. Pattern Recogn. **37**(3), 487–501 (2004)
21. Calinski, T., Harabasz, J.: A dendrite method for cluster analysis. Commun. Stat. **3**(1), 1–27 (1974)
22. Halkidi, M., Vazirgiannis, M.: Clustering validity assessment: finding the optimal partitioning of a data set. In: Proceedings 2001 IEEE International Conference on Data Mining, San Jose, CA, USA, pp. 187–194 (2001)
23. Saitta, S., Raphael, B., Smith, I.F.C.: A bounded index for cluster validity. In: Proceedings of the 5th International Conference on Machine Learning and Data Mining in Pattern Recognition, Leipzig, Germany, pp. 174–187 (2007)
24. Wang, R., Lai, S., Wu, G., Xing, L., Wang, L., Ishibuchi, H.: Multi-clustering via evolutionary multi-objective optimization. Inf. Sci. **450**, 128–140 (2018)
25. Morimoto, C.Y., Pozo, A.T.R., de Souto, M.C.P.: An adaptive evolutionary multi-objective clustering based on the data properties of the base partitions. Expert Syst. Appl. **245**, 123102 (2024)
26. Garza-Fabre, M., Handl, J., José-García, A.: Evolutionary multiobjective clustering over multiple conflicting data views. IEEE Trans. Evol. Comput. **27**(4), 817–831 (2022)
27. Mohammadi, S., Desai, V., Karimipour, H.: Multivariate mutual information-based feature selection for cyber intrusion detection. In: IEEE Electrical Power and Energy Conference (EPEC), pp. 1–6. IEEE (2018)
28. Mukhopadhyay, A., Maulik, U., Bandyopadhyay, S.: Simultaneous informative gene selection and clustering through multiobjective optimization. In: IEEE Congress on Evolutionary Computation, pp. 1–8. IEEE (2010)

29. Zhao, F., Tang, Z., Xiao, Z., Liu, H., Fan, J., Li, L.: Ensemble CART surrogate-assisted automatic multi-objective rough fuzzy clustering algorithm for unsupervised image segmentation. In: Engineering Applications of Artificial Intelligence, vol. 133, article 108104, (2024). https://doi.org/10.1016/j.engappai.2024.108104
30. Mukhopadhyay, A., Ray, S., Maulik, U., Bandyopadhyay, S.: Multiobjective optimization algorithms for bioinformatics, Springer Singapore (2024). ISBN 978-981-97-1630-2. https://doi.org/10.1007/978-981-97-1631-9
31. Rodríguez-Bejarano, F.M., Hernández-Lobato, J.C., Rodríguez, M.I., Rubio, A.: Reference vector-guided evolutionary algorithm for cluster analysis of single-cell transcriptomes. Computer Methods and Programs in Biomedicine, vol. 242, article 108873 (2025). https://doi.org/10.1016/j.cmpb.2025.108873
32. Sánchez-Gómez, J.M., Santander-Jiménez, S., Vega-Rodríguez, M.A.: Multi-objective swarm-intelligence algorithm for document clustering. Expert. Syst. Appl., 289, Article 128348 (2025). https://doi.org/10.1016/j.eswa.2025.128348
33. Su, Y., Zhou, K., Zhang, X., Cheng, R., Zheng, C.: A parallel multi-objective evolutionary algorithm for community detection in large-scale complex networks. Inf. Sci. **576**, 374–392 (2021). https://doi.org/10.1016/j.ins.2021.06.089
34. Yuste, J., Pardo, E.G., Duarte, A.: Multi-objective variable neighborhood search for improving software modularity. In: Proceedings of the International Conference on Variable Neighborhood Search (VNS 2022), pp. 58–68. Springer Nature Switzerland, Cham (2022). https://doi.org/10.1007/978-3-031-19647-0-6
35. Korlakov, S., Klassen, G., Bauer, L.T., Conrad, S.: Multi-objective optimisation for the selection of clusterings across time. Eng. Proc. **68**(1), article 48 (2024). https://doi.org/10.3390/engproc2024068048

GPSR Compliance
The European Union's (EU) General Product Safety Regulation (GPSR) is a set of rules that requires consumer products to be safe and our obligations to ensure this.

If you have any concerns about our products, you can contact us on

ProductSafety@springernature.com

In case Publisher is established outside the EU, the EU authorized representative is:

Springer Nature Customer Service Center GmbH
Europaplatz 3
69115 Heidelberg, Germany

www.ingramcontent.com/pod-product-compliance
Ingram Content Group UK Ltd.
Pitfield, Milton Keynes, MK11 3LW, UK
UKHW021614260926
475829UK00010B/44
9789819579075